Encyclopaedia of

SOLID STATE CHEMISTRY

Encyclopaedia of
SOLID STATE CHEMISTRY

C.K. Choudhary

ANMOL PUBLICATIONS PVT. LTD.
NEW DELHI-110 002 (INDIA)

ANMOL PUBLICATIONS PVT. LTD.
Regd. Office: 4360/4, Ansari Road, Daryaganj,
New Delhi-110002 (India)
Tel.: 23278000, 23261597, 23286875, 23255577
Fax: 91-11-23280289
Email: anmolpub@gmail.com
Visit us at: www.anmolpublications.com

Branch Office: No. 1015, Ist Main Road, BSK IIIrd Stage
IIIrd Phase, IIIrd Block, Bangalore-560 085 (India)
Tel.: 080-41723429 • Fax: 080-26723604
Email: anmolpublicationsbangalore@gmail.com

Encyclopaedia of Solid State Chemistry

First Edition, **2011**

ISBN 978-81-261-4244-6

PRINTED IN INDIA

Printed at Mehra Offset Press, Delhi.

Contents

Preface

Solid-state chemistry is the study of the synthesis, structure, and physical properties of solid materials. It therefore has a strong overlap with solid-state physics, mineralogy, crystallography, ceramics, metallurgy, thermodynamics, materials science and electronics with a focus on the synthesis of novel materials and their characterization.

Solid state inorganic chemistry has been strongly driven by technology well in advance of atomic-level descriptions or in the academic studies. This subject teaches basic principles of chemistry and shows how they apply in describing the behavior of the solid state. The relationship between electronic structure, chemical bonding, and crystal structure is developed.

The science of the elementary, atomic compositions of solids and the transformations that occur in and between solids and between solids and other phases to produce solids. Solid-state chemistry deals primarily with those microscopic features which are uniquely characteristic of solids and which are the causes for the macroscopic chemical properties and the chemical reactions of solids. As with other branches of the physical sciences, solid-state chemistry also includes related areas that furnish concepts and explanations of those phenomena which are more characteristic of the subject itself.

The present Encyclopaedia explains the way metals and molecules that bind as ligands interact, and the consequences of this assembly process. It describes the chemical and physical properties and behaviour of the complex assemblies that form, and applications that may arise as a result of these properties.

Author

Chapter 1

Introduction

ATOMS

The familiar model of an atom is that of a small nucleus composed of protons and neutrons surrounded by rapidly moving electrons. Typically, the atomic diameter is on the order of 10^{-10} m while that of the nucleus is on the order of 10^{-15} m. Protons and neutrons have about the same mass (1.00728 and 1.00867 amu respectively) and each is about 1800 times as heavy as an electron.

A neutron is electrically neutral, but a proton has a positive charge ($+1.6 \times 10^{-19}$ coulomb) which is exactly the opposite of the negative charge of an electron. In a neutral atom, the number of electrons around the nucleus equals the number of protons in the nucleus.

The number of protons in the nucleus (the "atomic number", Z) characterizes a chemical element. Atoms of a given element all have the same number of protons, yet may have different masses. The atomic mass number of an atom, A, is given by $A = Z + N$, where N is the number of neutrons in the nucleus.

Since an element is characterized solely by Z, it follows that atoms of a given chemical element may have a varying number of neutrons. Subspecies of chemical elements with the same Z but differing N and A are called isotopes. The atomic weight of an element is the weighted average of the atomic masses of the various naturally occurring isotopes of the element, and the atomic weight scale is based on a value of exactly 12, after the carbon isotope that has an atomic mass number of 12.

NUCLEI

The nucleus of an atom weighs less than the sum of the weights of its isolated component particles. The difference between the actual mass and that of the components is called the mass defect. The mass defect, Δ m, is related to the binding energy within the nucleus, Δ E (in Joules), through Einstein's equation:

$$\Delta E = \Delta m c^2$$

where c is the velocity of light (in $m.s^{-1}$) and Δm is the mass defect (in kg). The nuclear forces which bind protons and neutrons together are strong, and the binding energy per nuclear particle (nucleon) amounts to about 1.4×10^{-12} Joules.

The greatest nuclear binding energy is found in nuclei of medium atomic number (such as Fe) where N is approximately equal to Z. For nuclei of larger atomic number, such as uranium, N is about equal to 1.5 Z, and the binding energy per nucleon is less.

As a consequence of this decreased nuclear stability, some isotopes (of uranium, for example) are unstable. That is, if the uranium isotope, ${}_{92}U^{235}$ (Z = 92, A = 235), is bombarded with neutrons, the following reaction can take place:

$${}_{92}U^{235} + n \rightarrow {}_{39}Y^{94} + {}_{53}I^{140} + 2n$$

Here the reaction products are smaller nuclei of increased stability. (In the above convention, subscripts indicate the atomic number and the superscripts the mass number.) Notice that one incident neutron generates fission products including two neutrons - the basis for chain reactions in nuclear reactors and nuclear explosions.

Heavy nuclei (even light nuclei) which have an unfavorable ratio in the number of protons and neutrons can spontaneously decay by the emission of α particles (helium ions) or β particles (electrons). These nuclei are referred to as radioactive. The rate at which the decay of such unstable nuclei takes place varies greatly and is indicated by the half-life of the material.

In one half-life period, half of the unstable nuclei will have emitted radiation and thus will have changed their character (atomic number). In two half-life periods, only 1/4 of the nuclei will have survived. In three half-lives, only 1/8 of the original nuclei remain, etc.

For example, the half-life of gamma-emitting "radio" cobalt, ${}_{27}Co^{60}$ (used for X-ray therapy), is 5.3 years, whereas that of radioactive ${}_{6}C^{14}$ is 5700 years. (Much more radiation is emitted per second by a given number of Co^{60} atoms than by the same number of C^{14} atoms.)

EXTRA-NUCLEAR ELECTRONS IN ATOMS

The first atomic theory in quantitative agreement with some experimentally determined facts was proposed in 1913 by Niels Bohr. He postulated (for atomic hydrogen) that:

- The (extra-nuclear) electron can assume only distinct (quantized) energy levels or states.
- In such energy levels, the electrons in motion will not radiate

(loose) energy; on changing energy levels, radiation equivalent to the energy difference between the levels is involved.

$$E_{rad} = E_2 - E_1 = h\upsilon$$

- The stable states of the atom involve motion of the electrons in circular orbits.
- The angular momentum of electrons in orbit (mvr) is an integral number of h/2.

$$mvr = n\frac{n}{2\pi}$$

- Newtonian mechanics applies to orbiting electrons.

A pictoral view of the Bohr atomic model visualizes electrons orbiting at a velocity (v) in well-defined spherical orbitals of radius (r) around the nucleus. The angular momentum of the orbiting electrons can, as stated above, only assume certain values of $n \times h/2r\pi$ where n (called the principal quantum number) can assume any positive integer value, i.e.1, 2, 3, 4, etc. to infinity.

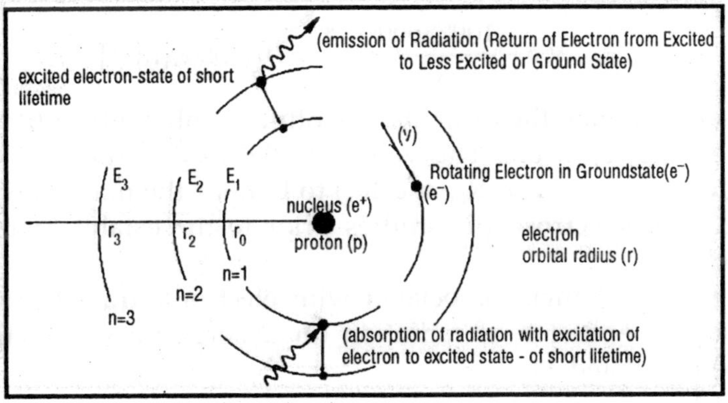

Fig. The Hydrogen Atom According to Bohr.

THE HYDROGEN ATOM

Assuming coulombic interaction and the applicability of Newtonian mechanics, the following values may be directly obtained for the hydrogen atom:

- The radius of an electron orbit (spherical) in hydrogen is given by:

$$r_n = \frac{n^2h^2\varepsilon_o}{\pi me}(n = 1,2,3,4,\text{etc.})$$

$$r_n = n^2 \times \text{constant} = n^2(0.529 \times 10^{-10})\text{m}$$

It can be seen that the smallest electron orbit in hydrogen - the stable orbit - is given for n = 1 at $r_o = 0.529$ Å. It is also seen that r increases with n^2 ($r = n^2 r_o$).

- The energy of an orbiting electron characterized by the principal quantum number (n) is given as:

$$E_n = -\frac{1}{n^2} \times \frac{me^4}{8h^2\varepsilon_0^2} = \frac{1}{n^2} \times \text{Constant[Joule]}$$

This relationship indicates that electron orbits with increasing principal quantum number (n) assume decreasing negative values ($1/n^2$) with the limiting value (n∞) being zero. For convenience, it is frequently customary to express electronic energies in terms of wave numbers ($\bar{v}$) with the units of [m–1]. This conversion may readily be made since E = h. Thus:

$$\bar{V} = -\frac{1}{\lambda} = -\frac{1}{n^2} \times \frac{me^4}{8h^3c\varepsilon_0^2}[m^{-1}]$$

Upon substituting numerical values for all the constants, we obtain:

$$\bar{V} = -\frac{10973500}{n^2} = \frac{1}{n^2} \times 10973500[m^{-1}]$$

(where n is again the principal quantum number which may assume the values 1, 2, 3, 4, etc. The value of 10973500 is referred to as the Rydberg constant (R; also R_H and R_∞) to honour the man who obtained its value from spectroscopic studies prior to the establishment of the Bohr model.)

The energy changes associated with electronic transitions and the accompanying emission of radiation (or absorption of radiation) may simply be calculated as

Follows:

$$\text{Emission} = \Delta E = (E_{n_1} - E_{n_2}) = hv$$

Here n2 is the principal quantum number of the outer orbit and n1 is that of the inner orbit. In terms of ?, the wave number, we obtain for emission of radiation:

$$\bar{v} = \left[\left(-\frac{10973500}{n_1^2}\right) - \left(-\frac{10973500}{n_2^2}\right)\right]$$

or

$$\Delta E = 10973500\left(\frac{1}{n_2^2} \quad \frac{1}{n_1^2}\right)$$

$$\Delta E = R\left(\frac{1}{n_2^2} - \frac{1}{n_1^2}\right) \times hc$$

$$\Delta E = 2.1798 \times 10^{-18}\left(\frac{1}{n_2^2} - \frac{1}{n_1^2}\right)J$$

$$\Delta E = En_2 - En_1$$

$$\bar{v} = R\left(\frac{1}{n_1^2} - \frac{1}{n_2^2}\right)$$

Electronic transitions are conveniently represented in so-called term schemes.

MULTI-ELECTRON SYSTEMS

The limitations of the Bohr atom became apparent soon after its establishment. For example, the uranium atom would require 92 electronic orbits and it would have to assume a size which is incompatible with the experimental facts.

The first modifications of the Bohr model were made by Sommerfeld, who introduced additional quantum conditions: Taking into account that individual sharp spectral lines split into several lines (of slightly differing λ) if light emission takes place in a magnetic field (Zeeman effect), the following picture emerged from the Bohr atom: electronic spherical orbits (shells)

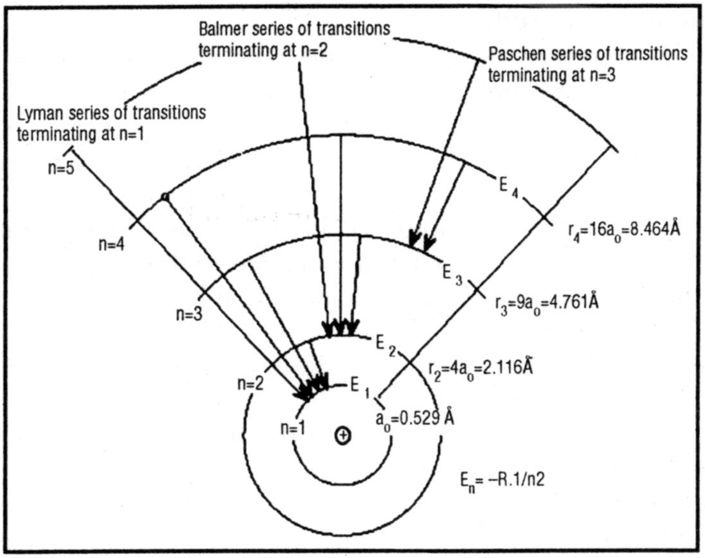

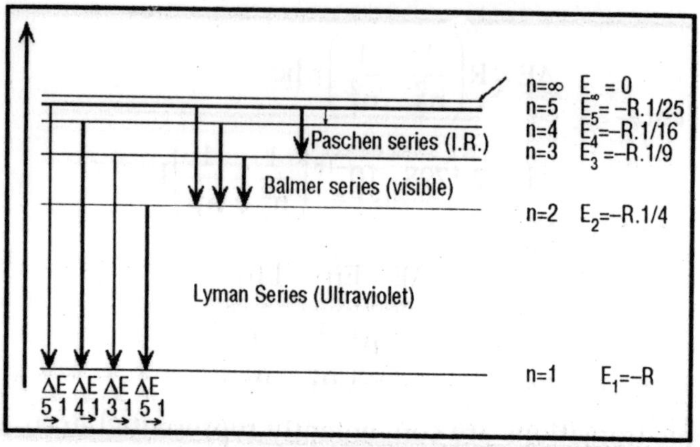

Fig. Electron Transitions and Related Optical Spectra in Atomic Hydrogen.

may have suborbitals (orbits) which are elliptical of varying eccentricity - controlled by an additional quantum number, the orbital quantum number (l).

The eccentric orbitals may assume only particular spatial orientations controlled by the magnetic number (m), in an external magnetic field. The observed magnetic behaviour of the elements required an additional quantization (Uhlenbeck) – the introduction of the spin quantum number (s).

According to the Bohr–Sommerfeld quantum theory, electrons move about the nucleus of an atom in well-defined orbits, each of which is characterized by four numbers, called quantum numbers. An electron moving in an orbit close to the nucleus has a larger (negative) energy than one in an outer orbit.

Energy is therefore necessary to move an electron from an inner to an outer orbit (with smaller negative energy). Conversely, energy is released (as radiation) during the transition of an electron from an outer to an inner orbit. The (electromagnetic) radiation released consists of discrete entities or quanta, which are called photons.

The energy of emitted-photons is given by the difference in energy between the two orbits involved and is related to the frequency, v, or the wavelength, λ, of the electromagnetic radiation by:

$$E_{Photon} = E_2 - E_1 = hv = h(c/\lambda) \qquad \{c = \lambda v\}$$

where h is called the Planck's constant, E_2 and E_1 are the energies of the outer and inner orbits, respectively, and c is the velocity of light. Through (spectroscopic) observations of the frequencies or wavelengths of radiation emitted by an excited atom (atom with an excited electron), the differences of electron energy levels can be determined. Figure shows such energy levels obtained from measurements of the wavelengths of radiation emitted by atomic hydrogen excited electrically in a gas discharge tube.

To specify the spatial location and energies of electrons in an atom requires the use of four quantum numbers. The "principal quantum number", (n), determines the size of the shell in which a particular electron orbits around the nucleus; it also largely determines its energy. The value of n is restricted to integral values: n = 1, 2, 3 and so forth. A value of n = 1 signifies that the electron exists in the lowest energy state and its orbit is in the innermost allowed shell, as close to the nucleus as possible.

Successively higher energy states are represented by n = 2, 3, etc. It should thus be clear that the principal quantum number of the outermost orbiting electron determines, in principle, the size of a given atom. For convenience, letters are frequently used to characterize the electronic shells of n = 1, 2, 3, 4, etc. In letter notation:

n = 1 = K shell

n = 2 = L shell

n = 3 = M shell

etc.

The second quantum number, (l), is referred to as the "orbital" or "azimuthal quantum number". It specifies the angular momentum of orbiting electrons and, to a minor extent, their energy. "l" can only assume integral values ranging from l = 0 to l = (n–1) (where n is the principal quantum number).

Accordingly, for n = 1 (the K shell), l can only assume a value of 0. For n = 2, either l = 0 or l = 1 is possible, etc. For convenience again, letters are used to specify the electronic state (orbit) in a given shell (characterized by n) corresponding to the second quantum number:

l = 0 = s (orbit)

l = 1 = p (orbit)

l = 2 = d (orbit)

l = 3 = f (orbit)

An electron with l = 0 is in an s–quantum state (in short, it is an s-electron), one with l = 1 is in a p-state, etc. Thus, an electron with a principal quantum number n = 3 and an orbital quantum number l = 1 is called a 3p electron (the electron in the M shell is in a p orbital).

The third quantum number, (m), called the "magnetic quantum number", controls the number of allowed spatial orientations ("degeneracy") of each orbit characterized by l in a given shell [characterized by (n)].

Degenerate states are of identical energy. The "degeneracy" disappears in the presence of a magnetic field where different spatial orientation of orbits assume different energy values. The total number of allowed orbital orientations for any orbital characterized by l is (2l + 1), corresponding to m values of l, (l – 1), (l – 2), ..., 0, –1, –2, ..., –l.

An electronic state with l = 0 necessarily has only m = 0 and thus has no directional orientation in space; it is a spherical orbit. On the other hand, a p-state (l = 1) allows m values of –1, 0, +1. The resulting three possible orientations are perpendicular to one another.

The first three quantum numbers (n, l and m) define atomic electron orbitals. They are related respectively to the size, shape, and spatial orientation of the orbital. The fourth quantum number, s, (called the "electron spin quantum number"), can be interpreted as determining for an orbiting electron the direction of electronic spin around its own axis. "s" can assume values of +1/2 (parallel spin) and –1/2 (anti-parallel spin).

For notational purposes, a positive s is conveniently represented as ↑and a negative s as ↓. We may now summarize our findings about the electronic states in atoms. Any electron in an atom is defined by four quantum numbers, viz.:

- The principal quantum number (n), which may have any positive integral value except zero. (In practice, because of the instability of heavy nuclei, atoms containing electrons with n values greater than 7 are unknown.)
- The orbital or azimuthal quantum number (l), which is an integer and may have any positive value less than n, including zero.

$$0 \leq l < n$$

- The magnetic quantum number (m), which is also an integer and may have any positive or negative value equal or less than l, including zero.

$$-l \leq m \leq +l$$

- The spin quantum number (s) which may only assume one of two values, namely +1/2 and –1/2.

$$S = \pm 1/2$$

An important law enables us to make use of these quantum rules for the characterization of electronic states in multi-electron systems: the Pauli Exclusion Principle. It states that in any atom no two electrons may have the same four quantum numbers.

From this principle it follows that each electronic orbital can accommodate at most two electrons differing by their spin quantum number which will be +1/2 for one electron and –1/2 for the other electron. (The Pauli principle is based on the fact that the separate existence of any electron depends upon its non-destruction by interference, i.e. on its wave nature.)

A further useful fact that generally simplifies understanding electronic structures is that (with exceptions stated shortly) the quantum states for electrons follow the rule (Aufbau Principle) that the lowest m, l and n numbers, consistent with Pauli's exclusion principle, are selected first by electrons in multi-electron atoms.

(For the spin quantum numbers, the +1/2 value is given priority over the –1/2 value.) Applicability of this rule is restricted to systems in which the orbitals, defined by a selected set of quantum numbers with lowest possible numerical value, correspond to orbitals of lowest energy since in all instances the lowest energy levels are filled first. The build-up of the electronic states of an atom is obtained by placement of the electrons first in the orbitals of lowest energy (the aufbau principle or "construction principle").

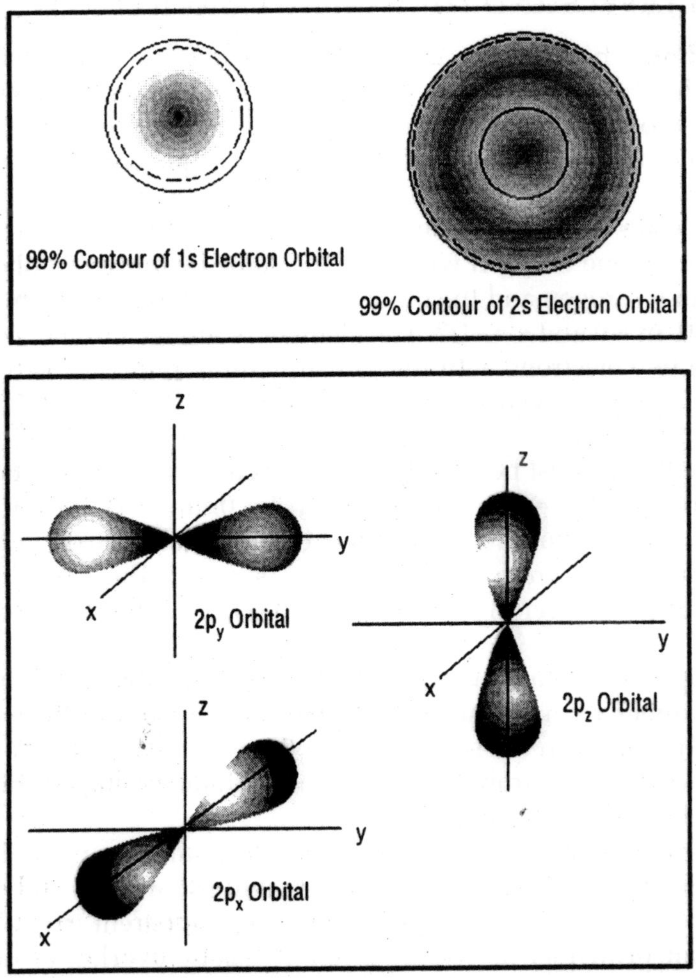

Fig. Conclusions on Electrion Orbital Shapes Based on Wave Moechanics

Wave mechanics (introduced by Schrödinger), unlike the quantum theory based on a planetary model, asserts that an electron in an atom cannot be considered as a particle having an orbit with a definite radius. Instead, there is a probability of an electron being at certain spatial positions.

Hence, the location of an electron is best described in terms of its probability density distribution, which is sometimes called an electron cloud. The spatial symmetry of the probability distribution depends upon the electronic state. The electron cloud is spherically symmetric for s–electrons, but more complicated for electrons in a p-state. Examples of these distributions are shown in figure for 1s and 2p electrons.

ENERGY LEVELS AND THE AUFBAU PRINCIPLE

Consider the atom of an element containing one extra-nuclear electron: hydrogen; for electro-neutrality the charge on the nucleus must be +1. This orbiting electron (in the ground state - the lowest possible, most stable state) will have the lowest available quantum numbers in n, l and m.

That is, n = 1; hence l = 0 and m = 0; also, s = +1/2. In the two-electron atom helium, one electron will have the same quantum numbers as the electron in hydrogen and the other electron will have quantum numbers n = 1, l = 0, m = 0 and s = –1/2. The element of atomic number 3 (lithium) will have two electrons with the same quantum numbers as the helium electrons, plus one electron with the quantum numbers n = 2, l = 0, m = 0, s = +1/2.

[Note that n must be 2 for the third electron since states such as (n = 1, l = 1) or (n = 1, l = 0, m = 1) or (n= 1, l = 0, m = –1) are not allowed. Similarly, the quantum numbers for the additional electron in beryllium (atomic number 4) are n = 2, l = 0, m = 0, s = –1/2. The next electron for boron (atomic number 5) has the values n = 2, l = 1, m = –1, s = +1/2, and so on.

In applying the aufbau principle to the orbital filling with increasing atomic number we have thus far inferred the tendency of the energies of electrons to follow in the same order as the principal quantum numbers; that is, successive electron shells are filled with increasing Z. This concept does not hold in all instances.

For example, electrons with the quantum numbers n = 3, l = 2 (and various m values) are of higher energy than those with n = 4, l = 0. Other inversions occur with n = 4 and higher. These apparent irregularities in the aufbau principle, the result of an energetic overlap of orbitals in successive shells, lead to partial shell fillings and the appearance of groups of so-called transition elements.

The groups are characterized (with minor exceptions) by identical outermost electron shell configurations and therefore do not exhibit, with increasing atomic number, the (expected) change in properties observed on regular shell filling for elements 2–18. (In the transition elements, with increasing atomic number the electrons are accommodated in lower lying shells which remained empty because of the above mentioned energy overlap of their orbitals.) The energy levels of the various orbitals in shells with increasing n are schematically indicated.

Another factor in the application of the aufbau principle is Hund's Rule which states that in atoms the electrons tend at first to fill up given orbitals (m levels) singly (with unpaired spins and spin quantum number equal to +1/2).

Only after all m levels associated with a particular l value in a given shell have been used for single-electron occupation does doubling of electrons into m levels occur.

Thus the extra electron for carbon (atomic number 6) has the quantum numbers n = 2, l = 1, m = 0, s = +1/2 rather than n = 2, l = 1, m = –1, s = –1/2. Both the effects of energy inversion and Hund's rule are very apparent when considering the electronic configuration of the elements Z = 18, 19, 20 and 21.

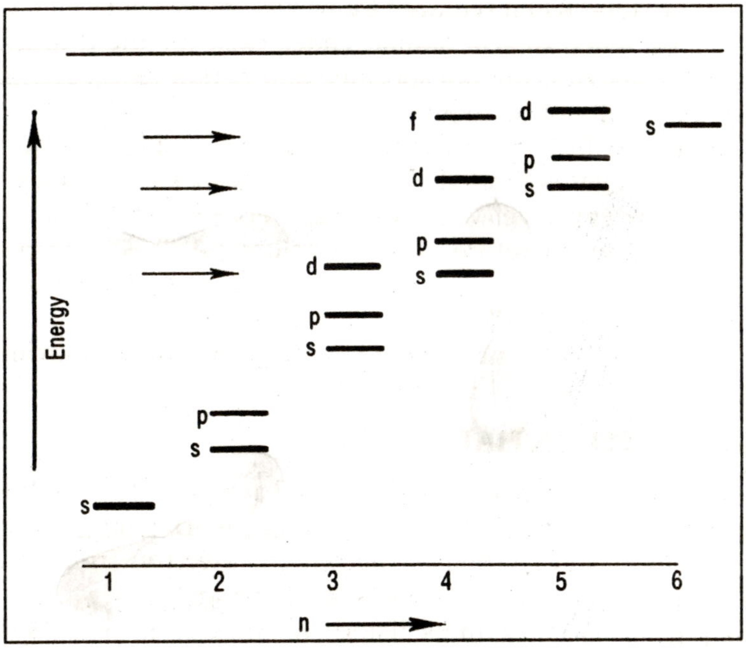

Fig. Relative Energies of the Orbitals in Neutral, Many-electron Atoms.

Electrons will always assume lowest available energy states.

Accordingly, with increasing atomic numbers, the 4s states will be filled prior to the 3d states, for example.

The s, p, d, and f orbital sets:

Types of orbitals	Orbital quantum numbers	Total number of orbitals in set	Total number electrons accommodated.
s	1 = 1; m=0	1	2
p	1=1;m=1,0–1	3	3
d	1=2;m=1,0,–1,–2	5	10
f	1=3; m=3,2,1,0,-1,-2,-3	7	14

The presently used notation is not the most convenient way of designating the energy levels (and wave functions) occupied by the electrons in atoms. The more informative way is a code written as follows: The principal quantum number is given first, followed by the conventional letter which designates the azimuthal (orbital) quantum number and, in a superscript, the number of electrons with that azimuthal quantum number.

The code for the electronic ground state in hydrogen, therefore, is 1s1, for helium $1s^2$, lithium $1s^{22}s^1$, beryllium $1s^{22}s^2$, boron $1s^{22}s^{22}p^1$, etc. You will note that the magnetic quantum number has not been specified, but this can be done when required by writing px when m = –1, py when m = +1 and pz when m = 0. Similar codes exist for the d and f levels. Usually such subscripts are not specified and in that case a consideration of Hund's rule will not be required.

An important fact to remember is that the number of electrons in any s, p, d or f level is limited: There are, as you know, a maximum of two electrons in any level with spin numbers +1/2 and –1/2. For the p levels, l is one and, therefore, three m values (–1, 0, +1) are permitted, each with two electrons of spin number +1/2 and –1/2, resulting in a maximum of six p electrons in any p orbital system.

The reader should confirm that ten electrons are the maximum in a d system of orbitals and 14 in an f system of orbitals.

IONIZATION POTENTIAL

Electrons may be removed from isolated atoms by bombardment with other electrons and by heat, for examples. The work (energy) required to remove the most weakly bound (outermost) electron from an isolated atom is known as the "ionization energy".

This energy is sometimes listed in units of Joules. More often ionization energies are given in terms of the "ionization potential". The ionization potential is the potential (V) that will accelerate an electron at rest so that it acquires a kinetic energy, sufficient to extract the outermost

(most loosely bound) electron from an atom. This potential is 13.595 Volt (as listed in the PT) for the ionization of a hydrogen atom.

You can also say the electron is bound to the proton by an (negative) energy equivalent to the energy of an electron accelerated by a potential of 13.595 Volt.

[If you make a dimensional analysis, you will recognize that charge (e) times potential (V) has the dimensions of energy ($kg.m^2.sec^{-2}$) and 1 electron Volt (1eV) corresponds to 1.6 x 10^{-19} Joules.]

It is significant that inert or noble gases have some of the highest ionization potentials. This reflects the fact that these elements have just enough electrons to completely fill a shell or a subshell which form stable configurations.

On the other hand, the ionization potentials of alkali metal atoms (Li, Ni, etc.) are low - the lowest, that of Cs, is only 3.89 eV. The reason for this is that alkali metal atoms have one outer s-electron beyond the stable electronic structure of an inert gas atom. Consequently this single electron in

Table. Atom Atomic Electron Affinities (EA)

Atom	Orbital electronic	EA electron Configuration	Orbital electronic Volts configuration of anion
H	$1s^1$	0.756	(He)
F	(He) $2s^2$2p5	3.45	(Ne)
Cl	(Ne) $3s^2$3p5	3.61	(Ar)
Br	(Ar) $4s^2 3d^{10} 4p^5$	3.36	(Kr)
I	(Kr) $5s^2 4d^{10} 5p^5$	3.06	(Xe)
O	(He) $2s^2 2p^4$	1.47	(He)$2s^2 2p^5$
S	(Ne) $3s^2$3p4	2.07	(Ne)$3s^2$3p5
Se	(Ar) $4s^2 3d^{10} 4p^4$	(1.7)	(Ar)$4s^2 3d^{10} 4p^5$
Te	(Kr) $5s^2 4d^{10} 5p^4$	(2.2)	(Kr)$5s^2 4d^{10} 5p^5$
N	(He) $2s^2 2p^3$	(–0.1)	(He)$2s^2 2p^4$
P	(Ne) $3s^2$3p3	(0.78)	(Ne)$3s^2$2p4
As	(Ar) $4s^2 3d^{10} 4p^3$	(0.6)	(Ar)$4s^2 3d^{10} 4p^4$

The outermost new shell can be removed relatively easily. (The second and successive ionization potentials of such atoms are increasingly greater.) From a chemical and physical viewpoint, it is found that whether an atom is neutral or ionized as well as isolated or combined leads to distinctly different characteristics. It should be recognized that the notation used previously for neutral atoms can also be employed for electronic structures of ions.

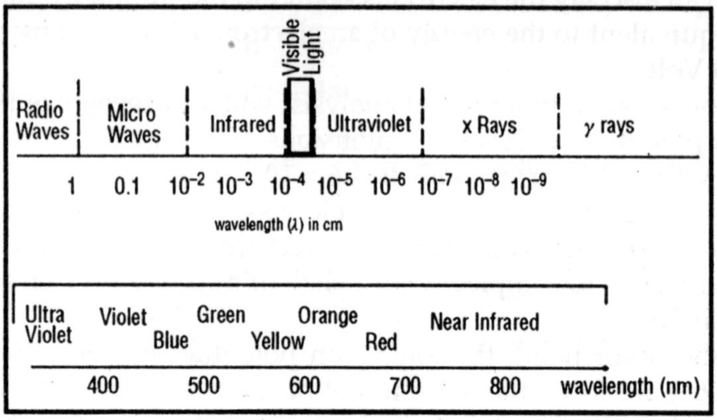

Fig. Wavelength.

The electronic configurations of an iron atom and two iron ions are shown below.

Fe:	$1s^2 2s^2 2p^6 3s^2 3p^6 3d^6 4s^2$	(neutral atom)
Fe^{+2}:	$1s^2 2s^2 2p^6 3s^2 3p^6 3d^6$	(ferrous ion)
Fe^{+3}:	$1s^2 2s^2 2p^6 3s^2 3p^6 3d^5$	(ferric ion)

ELECTRON AFFINITY

The tendency of some atoms to accept, as well as to lose, electrons is important in determining how atoms combine with each other. Whereas removal of an electron always requires the expenditure of energy, acceptance of one extra electron by an atom generally is accompanied by the release of energy. The amount of energy released on acquisition of an electron is called the electron affinity (or "first electron affinity").

The elements on the left of the periodic table (metals) and the inert gases on the outer right have low electron affinities, whereas the non-metals have higher ones. The particularly high electron affinities of halogen atoms (F, Cl, etc.) may be attributed to the fact that these elements all lack one electron which would provide the stable electronic structure of an inert gas. By acquiring an additional electron, they increase their stability since the additional electron goes into the orbital lacking one electron and thus results in a stable, s^2p^6, octet configuration.

ATOMIC SIZE

The size, or volume, of an isolated atom is difficult to define explicitly since, in the electron cloud model of the atom, the probability density distribution theoretically reaches zero only at infinity. Nevertheless, the electron density falls off so rapidly at a short distance from the nucleus that some approximation of size can be made. In the hydrogen atom, for example, the electron density is very nearly zero at a distance of 1.2 Å (1.2×10^{-10} m) from the nucleus.

The problem of defining atomic size is simplified in molecules and solids in that rather precise dimensions can be determined from interatomic distances which can be measured by diffraction techniques. Thus, in the H_2 molecule the atoms are only 0.72 Å apart, as determined from the distance between the nuclei. In this case the radius of the hydrogen atoms is taken to be 0.37 Å, even though the size of the H2 molecule is considerably more than four times this value.

It should be apparent that the atomic radius depends upon whether an atom is isolated or combined with other atoms. The radius of an isolated atom is called the van der Waals radius, that of a bound atom in a molecule is the covalent radius, and that of a bound atom in a metal is the metallic radius.

Van der Waals, covalent and metallic radii for some elements are listed in some periodic tables of the elements. The radii of positive ions (cations) and those of negative ions (anions) differ from the van der Waals radii.

SAMPLE PROBLEM

For hydrogen, calculate the radius (r) and energy level (ET) of the electron in the lowest energy state (ground state).

Solution

The size of the first allowed orbit can be calculated by requiring that the centrifugal force of the orbiting electron be balanced by the coulombic attraction to the nucleus.

$$F_{centr} = \frac{mv^2}{r} F_{att} = \frac{e^2}{4\pi\varepsilon_o r^2} \frac{mv^2}{r} = \frac{e^2}{4\pi\varepsilon_o r^2}$$

According to Bohr, the angular momentum of the elecrton (mvr) is quantized in units of $h/2\pi$:

$$mvr = \frac{nh^2}{2\pi} n = 1, 2, 3, \ldots$$

With these conditions, r for the first allowed orbit becomes:

$$\frac{mvr^2}{r} = \frac{e^2}{4\pi\varepsilon_0 r^2}$$

$$m^2v^2r^2 = \frac{me^2r}{4\pi\varepsilon_o}$$

$$\frac{n^2h^2}{4\pi^2} = \frac{me^2r}{4\pi\varepsilon_o}$$

$$r = \frac{n^2h^2\varepsilon_o}{\pi me^2} = n^2 \times \text{cons tan d}$$

For n = 1:

$$r_0 = 0.529 \times 10^{-10}\,\text{m} = 0.529\ \text{Å}$$

The kinetic energy of the electron is 1/2 mv^2. The potential energy as a function of distance from the nucleus is given by Coulomb's law as:

$$-\frac{e^2}{4\pi\varepsilon_o r}$$

The total energy is the sum of the kinetic and potential energies.

$E_T = E_K + E_P$

$$E_T = \frac{1}{2}mv^2 - \frac{e^2}{4\pi\varepsilon_o r}$$

The balance of attractive and repulsive forces requires that:

$$\text{Fcentr} = F_{att}\ \text{Since}: \frac{mv^2}{r} = \frac{e^2}{4\pi\varepsilon_o r^2}$$

$$\frac{mv^2}{r} = \frac{e^2}{4\pi\varepsilon_o r^2}\ \text{Then}: \frac{mv^2}{2} = \frac{e^2}{8\pi\varepsilon_o r}$$

Therefore:

$$E_T = \frac{e^2}{8\pi\varepsilon_o r} - \frac{e^2}{4\pi\varepsilon_o r} = -\frac{e^2}{8\pi\varepsilon_0 r}$$

Chpter 2

Nature of Crystalline Solids

In an assembly of atoms or molecules a solid phase is formed whenever the interatomic (intermolecular) attractive forces significantly exceed the disruptive thermal forces and thus restrict the mobility of atoms, forcing them into more-or-less fixed positions.

From energy considerations it is evident (as discussed in LN-2) that in such solids the atoms or molecules will always attempt to assume highly ordered structures which are characterized by symmetry. Depending on the nature of the acting interatomic forces, all solids may be subdivided into:

- Ionic solids (NaCl)
- Covalent solids (Diamond)
- Metallic solids (Fe, Ni, etc.)
- Van der Waals solids (ice, solid He)

Solids as we encounter them in nature may or may not reflect the internally ordered arrangement in their appearance. We find, for example, well-formed quartz crystals, garnets, diamonds and snowflakes which are all characterized by flat bounding planes which intersect at characteristic angles. On the other hand, we also observe rounded stones and man-made cast solid objects with no external evidence of internal order.

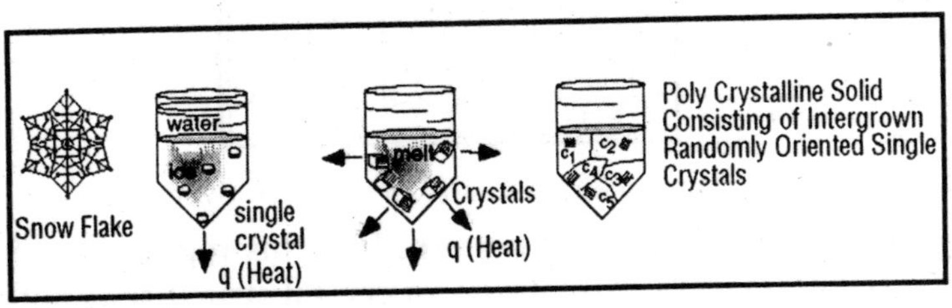

Fig. Crystals, internal Order and External Appearance.

To understand the external appearance of the solid state it is necessary to consider the formation of solids from different phases. Solids, for example, are formed upon cooling of liquids (melts) - by freezing or solidification; this solidification process normally proceeds in total confinement and the resulting "cast" structure will have an external appearance which reflects in detail the confining geometry (and not the internal order).

Moreover, depending on solidification conditions, the solid body may be either a single crystal or polycrystalline. Polycrystalline solids (in excess of 95% of the solid state encountered) may be thought of as an assembly of microscopic single crystals with random orientation held together like a maze structure by the interwoven irregular shapes of the individual crystals.

A typical example of an "unconfined" phase transformation is the formation of snowflakes where the external boundaries of the solid have assumed crystalline appearance, reflecting in detail the internally ordered molecular (H_2O) arrangement. Another unconfined formation of a solid is precipitation from solution (sugar crystals, $CuSO_4$ and the like). Similarly, the formation of crystals from the vapour phase leads to bodies which externally reflect elements of internal order.

CRYSTAL STRUCTURE

From the earlier discussion it should be apparent that, when strong interatomic forces exist, atoms tend to pack closely together - the closeness of packing being particularly pronounced in the solid state. In this case atoms can be regarded as hard spheres and the problem of close packing can be treated as one in which the whole assembly has a tendency toward efficient packing.

A little thought or a few simple experiments with ping-pong balls quickly convince us that regular arrangements of the spheres generally lead to more compact assemblies than irregular arrangements). The same principle applies to arrangement of atoms in the solid state. Where strong attractive forces are exerted we find that the atoms or molecules concerned arrange themselves

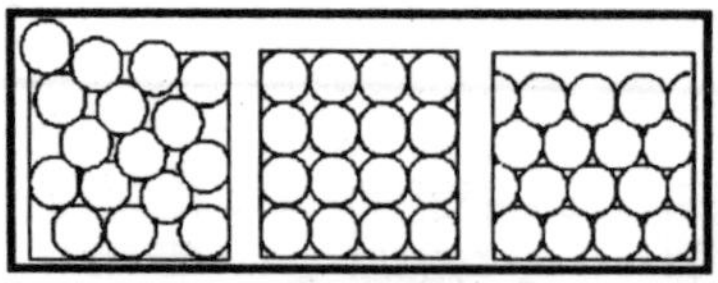

Fig. The Packing Density of Ordered Systems.

in a regular three-dimensional pattern. It is this regularity which is the basis of crystallinity in materials: i.e., a crystal structure is nothing more than an orderly array of atoms or molecules. This definition of a crystal is distinct from the popular concept based on observation of external symmetry of crystals, often seen during the study of elementary chemistry, in which some crystals appear cubic, others needle-shaped and so on. The regular external shape is obtained only when the conditions of crystallization are favorable to development of flat, geometric faces. In most instances, particularly with metals, these conditions are absent, and the crystals have irregular surfaces even though the internal arrangement is perfectly geometric.

Atomic arrays in crystals are conveniently described with respect to a three-dimensional net of straight lines. Consider a lattice of lines dividing space into equal-sized prisms which stand side-by-side with all faces in contact, thereby filling all space and leaving no voids. The intersections of these lines are points of a space lattice, i.e., a geometrical abstraction which is useful as a reference in describing and correlating symmetry of actual crystals.

These lattice points are of fundamental importance in describing crystals for they may be the positions occupied by individual atoms in crystals or they may be points about which several atoms are clustered. Since prisms of many different shapes can be drawn through the points of a space lattice to partition it into cells, the manner in which the network of reference lines is drawn is arbitrary.

It is not necessary that the lines be drawn so that atoms lie only at corners of unit prisms. In fact, it is more convenient to describe some crystals with respect to prisms in which atoms lie at prism centers or at the centers of prism faces as well as at prism corners.

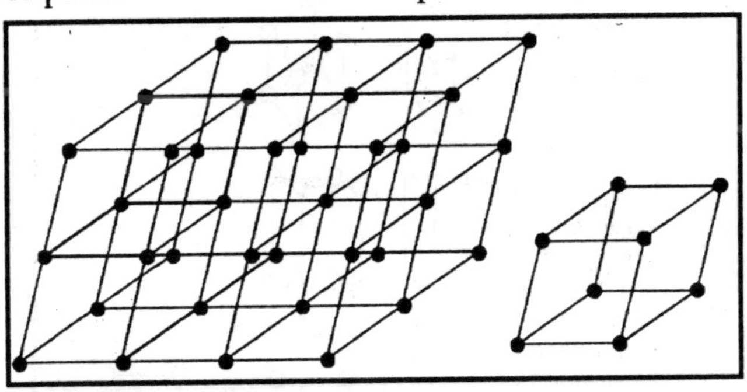

Fig. The Space Lattice Prisms in Which Atoms Lie at Prism Centers or at the Centers of Prism Faces as Well as at Prism Corners.

An important characteristic of a space lattice is that every point has identical surroundings: the grouping of lattice points about any given point is identical to the grouping about any other point in the lattice.

In other words, if we could move about in the lattice, we would not be able to distinguish one point from another because rows and planes near each point would be identical.

If we were to wander among the atoms of a solid metal or chemical compound, we would find the view from any lattice point exactly the same as that from any other.

There are fourteen space lattices. That is, no more than fourteen ways can be found in which points can be arranged in space so that each point has identical surroundings. Of course, there are many more than fourteen ways in which atoms can be arranged in actual crystals; thus there are a great number of crystal structures.

Too often the term "lattice" is loosely used as a synonym for "structure", an incorrect practice which is frequently confusing. The distinction can be clearly seen if we

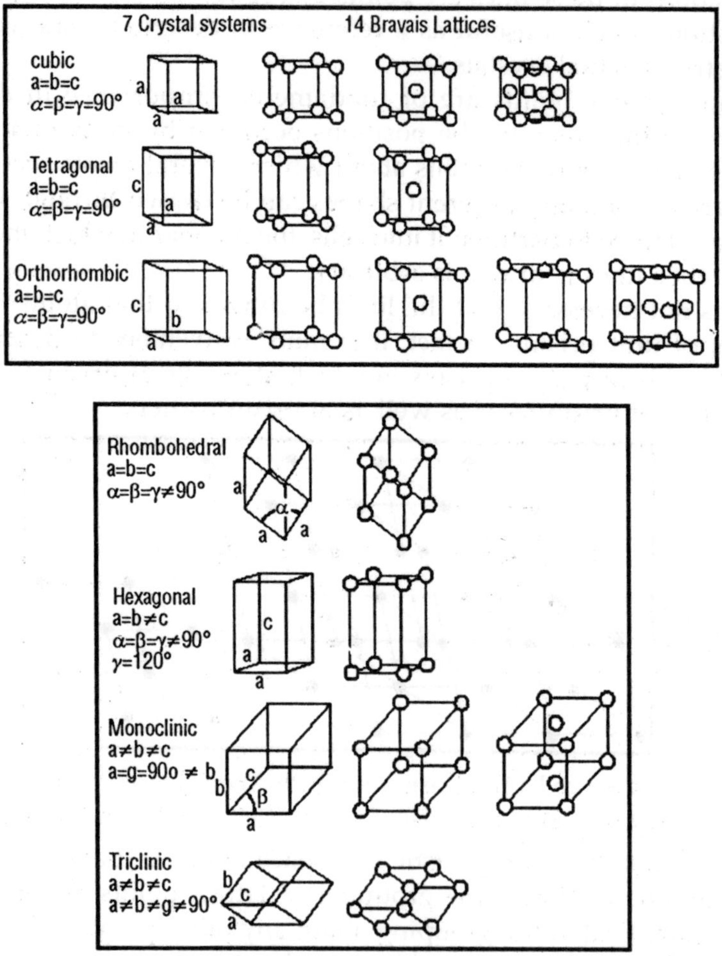

Fig. The 7 Crystal Systems and the 14 Bravais Lattices.

remember that a space lattice is an array of points in space. It is a geometrical abstraction which is useful only as a reference in describing and correlating symmetry of actual crystals. A crystal structure, however, is the arrangement of atoms or molecules which actually exists in a crystal. It is a dynamic, rather than a static, arrangement and is subject to many imperfections.

Although any crystal structure has an inherent symmetry which corresponds to one of the fourteen space lattices, one, two or several atoms or molecules in the crystal structure may be associated with each point of the space lattice. This symmetry can be maintained with an infinite number of different actual arrangements of atoms - making possible an endless number of crystal structures.

To specify a given arrangement of points in a space lattice, it is customary to identify a unit cell with a set of coordinate axes, chosen to have an origin at one of the lattice points. In a cubic lattice, for example, we choose three axes of equal length that

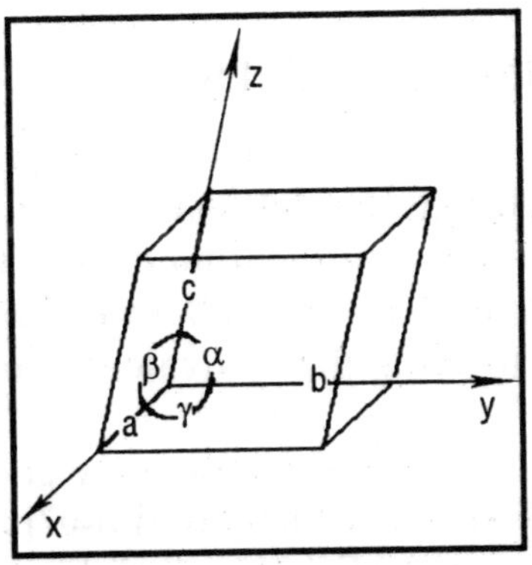

Fig. Specification of Unit Cell Parameters.

are mutually perpendicular and form three edges of a cube. Each space lattice has some convenient set of axes, but they are not necessarily equal in length or orthogonal.

Seven different systems of axes are used in crystallography, each possessing certain characteristics as to equality of angles and equality of lengths. These seven crystal systems are tabulated in Table (to be considered in conjunction with).

Table. The Seven Crystal Systems

System	Parameters	Interaxial Angles
Triclinic	$a \neq b \neq c$	$\alpha \neq \beta \neq \gamma$
Monoclinic	$a \neq b \neq c$	$\alpha = \gamma = 90° \neq \beta$
Orthorhombic	$a \neq b \neq c$	$\alpha = \beta = \gamma$
Tetragonal	$a = b \neq c$	$\alpha = \beta = \gamma$

Cubic	a = b = c	α = β = γ = 90°
Hexagonal	a = b ≠ c	α = β = 90°, γ = 120
Rhombohedral	a = b = c	α = β = γ ≠ 90°

The network of lines through the points of a space lattice divides it into unit cells. Each unit cell in a space lattice is identical in size, shape and orientation to every other unit cell. It is the building block from which the crystal is constructed by repetition in three dimensions. All crystal structures are based on these fourteen arrangements.

The body-centered cubic, face-centered cubic and hexagonal lattices are common and of prime importance in metals. Some of the metals associated with nuclear applications, such as uranium and plutonium, have crystal structures which are more complicated than these three relatively simple types. In general, crystalline ceramics also are more complex.

UNIT CELLS VS PRIMITIVE CELLS

In the literature we often find reference to unit cells and to primitive cells. The primitive cell may be defined as a geometrical shape which, when repeated indefinitely in three dimensions, will fill all space and is the equivalent of one atom. The unit cell differs from the primitive cell in that it is not restricted to being the equivalent of one atom. In some cases the two coincide.

For instance all fourteen space lattices are shown by their unit cells. Of these fourteen, only seven (which are those?) are also primitive cells. Primitive cells are drawn with lattice points at all corners, and each primitive cell contains the equivalent of one atom. For instance, a simple cubic unit cell has an atom at each corner.

However, at any of these given corners, this atom must be shared with seven other identical cubes which fill the volume surrounding this point. Thus there is effectively only 1/8 of the atom which can be assigned to that particular unit cell. Since there are eight corners in a cube, there is the equivalent of one atom, and thus the primitive cell and unit cell coincide. Continuing, consider the body-centered cubic (BCC) lattice. In this case there is one atom at the centre of the cube and one atom contributed by the eight corners. This cell, then, has two atoms and, to avoid confusion, should be termed a unit cell. In the face-centered cubic lattice there are six face atoms, but each face atom is shared by two cells.

Consequently, each face contributes 1/2 an atom. The faces thus contribute three atoms and the corners one, for a total of four atoms in the unit cell. The face-centered cubic structure (FCC) can also be considered as four interpenetrating simple cubic cells.

In the study of crystals the primitive cell has limited use because the unit cell more clearly demonstrates the symmetrical features of a lattice. In other words, the unit cell can usually be visualized readily whereas the primitive cell cannot. For example, the cubic nature of the face-centered cubic lattice is immediately apparent in the unit cell, but it is not nearly so obvious in the rhombohedral primitive cell.

PACKING OF ATOMS

A crystal structure is a regular array of atoms arranged on one of the fourteen space lattices. The least complicated crystal structures are those having a single atom at each lattice point. Polonium has the simplest structure, being simple cubic.

In normal metals, the atoms (or positive ions) are held together by a cloud of free electrons so that each atom tends to be attracted equally and indiscriminately to all its geometrically nearest neighbors by the free electrons passing between them. This condition fosters the formation of closely packed structures of the types which can be demonstrated by efficiently packing uniformly-sized spheres into a given volume.

You can gain a better grasp of packing by conducting an experiment as follows. Assume we have a quantity of small spheres which we are required to efficiently pack into a box (a 'two-dimensional' approach can be made using a handful of pennies).

After some shuffling, it is obvious that the closest possible packing is obtained when the spheres are in contact and their centers occupy positions which correspond to the apices of equilateral triangles. It is also evident that there are two sets of

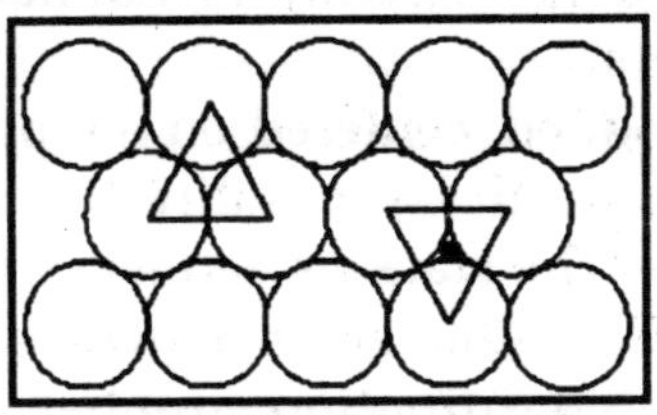

Fig. Close Packed Atomic Arrangements.

triangles – one set with vertices pointing away from the observer (points up) and the other set with vertices pointing toward the observer (points down).

When a second layer is added, there is closest-packing if the spheres in this new layer rest in the hollows formed by the spheres of the first layer. The centers of the spheres in the second layer will lie above the centers of the points-up triangles or above the centers of the points-down triangles, but not both simultaneously. Which set is used is immaterial.

For our discussion, however, assume the second layer is centered on the points-up triangles.

When we start adding a third layer, the spheres will again rest in the hollows formed by the spheres in the second layer. And again we have the option of placing the third layer on the points-up or on the points-down triangles. If we centre the third layer on the points-down triangles, we find the third layer is directly above the first layer.

If additional layers are added using an alternate stacking sequence (i.e., alternately centering the layers on the points-up and the points-down triangles), the sequence can be written as ABABABAB... This arrangement of spheres, translated to an arrangement of atoms, is the hexagonal close-packed (HCP) structure - very important, but not discussed in detail here.

Many elements having covalent bonding form arrangements in which the coordination number is (8–N), where N is the number of valence electrons. What, then, is the coordination number of the HCP structure just demonstrated? The geometry of the structure shows that any one atom has twelve equidistant neighbors. It is apparent that if, in any layer, a given sphere is placed in the adjoining layer, it fits in the hollow formed by three spheres and consequently is tangent to three spheres in the adjoining layer. Thus any given atom in the HCP structure is tangent to twelve other atoms – six in its own layer and three each in two adjoining layers.

When the third layer of spheres was added in the above discussion, we assumed this layer was centered on the points-down triangles of the second layer. What happens if the centers of the points-up triangles of the second layer are used instead? The distribution of spheres in the third layer is the same as in the first two layers, but does not lie directly above either of these two layers.

If a fourth layer is added, centered on the points-up triangles of the third layer, we find the fourth layer is directly above the first layer and duplicates it completely. The stacking sequence for this structure can be written as ABCABCABC... This arrangement has the same density of packing and the same coordination number as the HCP structure. However, it is the face-centered cubic (FCC) structure.

The HCP and FCC crystal structures have the same density of packing and the same coordination number. Therefore we might expect the behaviour of the two HCP and FCC structures to be very much alike with regard to physical and mechanical properties. This, however, in most instances is not the case. The maximum density of packing is found only in the HCP and FCC crystal structures. Why the metallic bond does not always produce one or the other of these two densest arrangements of atoms is as yet subject of intensive studies.

The BCC unit cell contains two atoms, and the coordination number is eight. There is partial compensation for this in the fact that there are six next-nearest neighbors at distances only slightly greater than that of the eight nearest neighbors. Some characteristics of cubic structures are given in Table. An interesting example of the type of crystal structure obtained in covalently bonded elements which obey the (8–N) coordination number rule (where N is the number of valence electrons) is the diamond structure. This structure is found in carbon, germanium, silicon, and tin at low temperatures and in certain compounds. Each atom has four nearest neighbors – a configuration that is variously called diamond cubic, body-centered tetrahedral or tetrahedral cubic.

Table. Characteristics of Cubic Lattices.

	Simple	Body-Centered	Face-Centered
Unit Cell Volume	a^3	a^3	a^3
Lattice Points Per Cell	1	2	4
Nearest Neighbour Distance	a	$\frac{a\sqrt{3}}{2}$	$\frac{a}{\sqrt{2}}$
Number of Nearest Neighbors	6	8	12
Second Nearest Neighbour Distance	$a\sqrt{2}$	a	a
Number of Second Neighbors	12	6	6

Many compounds crystallize in variations of cubic forms. Rock salt (NaCl), for example, is typical of many oxides, fluorides, chlorides, hydrides and carbides. It is sometimes considered as simple cubic with alternate Na and Cl atoms on the cube corners. Actually the structure is two interpenetrating FCC lattices – one of Na and the other of Cl – and the corner of one is located at point 1/2, 0, 0 of the other.

Many other oxides, fluorides and some intermetallic compounds have the fluorite (CaF_2) structure. This is FCC with Ca at the cube corners and face centers and F at all quarter points along the cube diagonals.

LATTICE PLANES AND DIRECTIONS

It is desirable to have a system of notation for planes within a crystal or space lattice such that the system specifies orientation without giving position in space. Miller indices are used for this purpose. These indices are based on the intercepts of a plane with the three crystal axes – i.e., the three edges of the unit cell.

The intercepts are measured in terms of the edge lengths or dimensions of the unit cell which are the unit distances from the origin along the three axes. For instance, the plane that cuts the x-axis at a distance from the origin equal to one-half the x-dimension of the cell is said to have an x-intercept equal to 1/2, and if it cuts the y-axis at 1/2 the y-dimension of the cell, the y-intercept is 1/2, regardless of the relative magnitudes of the x- and y-dimensions. If a plane is parallel to an axis, it intercepts the axis at infinity.

To determine Miller indices (hkl) of a plane, we take the following steps:

- Find the intercepts on the three axes in multiples or fractions of the edge lengths along each axis.
- Determine the reciprocals of these numbers.
- Reduce the reciprocals to the three smallest integers having the same ratio as the reciprocals.
- Enclose these three integral numbers in parentheses, e.g., (hkl).

A cube has six equivalent faces. If we have a definite orientation and wish to discuss one specific plane of these six, it is possible to specify this plane by using the proper Miller indices. Parentheses are used around the Miller indices to signify a specific plane.

On the other hand, it is often advantageous to talk about planes of a "form" – i.e., a family of equivalent planes such as the six faces of a cube. To do this it is customary to use the Miller indices, but to enclose them in curly brackets (braces). Thus the set of cube faces can be represented as {100} in which

{100} = (100) + (010) + (001) + (100) + (010) + (001)

This notation thus provides a shorthand scheme to avoid writing the indices for all six cube faces.

The utility of the scheme is even more evident in the case of the (110) planes – i.e., the dodecahedral planes (in a cubic system), where

{110} = (110) + (101) + (011) + (110) + (101) + (011)
+ (110) + (101) + (011) + (110) + (101) + (011)

The equivalent form for the orthorhombic system is:

{110} = (110) + (110) + (110) + (110)

{101} = (101) + (101) + (101) + (101)

{011} = (011) + (011) + (011) + (011)

The octahedral planes for the cube are:

{111} = (111) + (111) + (111) + (111) + (111) +(111) + (111) + (111)

Direction indices are defined in a different manner. A line is constructed through the origin of the crystal axis in the direction under consideration and the coordinates of a point on the line are determined in multiples of lattice parameters of the unit cell. The indices of the direction are taken as the smallest integers proportional to these coordinates and are closed in square brackets. For example, suppose the coordinates are x = 3a, y = b and z = c/2, then the smallest integers proportional to these three numbers are 6, 2 and 1 and the line has a [621] direction.

As further examples, the x-axis has direction indices [100], the y-axis [010] and the z-axis [001]. A face diagonal of the xy face of the unit cell has direction indices [110], and a body diagonal of the cell has direction indices [111].

Negative indices occur if any of the coordinates are negative. For example, the -y axis has indices [010]. A full set of equivalent directions, i.e., directions of a form, are indicated by carets: <uvw>.

X-RAYS AND X-RAY DIFFRACTION

X-rays were discovered during the summer of 1895 by Wilhelm Röntgen at the University of Würtzburg (Germany). Röntgen was interested in the cathode rays (beams of electrons) developed in discharge tubes, but it is not clear exactly which aspects of cathode rays he intended to study.

By chance he noticed that a fluorescent screen (ZnS^+ Mn^{++}) lying on a table some distance from the discharge tube emitted a flash of light each time an electrical discharge was passed through the tube. Realizing that he had come upon something completely new, he devoted his energies to investigating the properties of the unknown ray "X" which produced this effect. The announcement of this discovery appeared in December 1895 as a concise ten page publication.

The announcement of the discovery of X-rays was received with great interest by the public. Röntgen himself prepared the first photographs of the bones in a living hand, and use of the radiation was quickly adopted in medicine. In the succeeding fifteen years, however, very few fundamental insights were gained into the nature of X-radiation.

There was some indication that the rays were waves, but the evidence was not clear-cut and could be interpreted in several ways. Then, at the University of Munich in 1912, Max von Laue performed one of the most significant experiments of modern physics. At his suggestion, Paul Knipping (who had just completed a doctoral thesis with Röntgen) and Walter Friedrich (a newly appointed assistant to Sommerfeld) directed a beam of X-rays at a crystal of copper sulfate and attempted to record the scattered beams on a photographic plate.

The first experiment was unsuccessful. The result of a second experiment was successful. They observed the presence of spots produced by diffracted X-ray beams grouped around a larger central spot where the incident X-ray beam struck the film. This experiment demonstrated conclusively that X-radiation consisted of waves and, further, that the crystals were composed of atoms arranged on a space lattice.

ORIGIN OF X-RAY SPECTRA

The interpretation of X–ray spectra according to the Bohr theory (LN-1) of electronic levels was first (and correctly) proposed by W. Kossel in 1920: the electrons in an atom are arranged in shells (K, L, M, N, corresponding to n = 1, 2, 3, 4,..., etc.). Theory predicts that the energy differences between successive shells increase with decreasing n and that the electron transition from n = 2 to n = 1 results in the emission of very energetic (short wavelength) radiation, while outer shell transitions

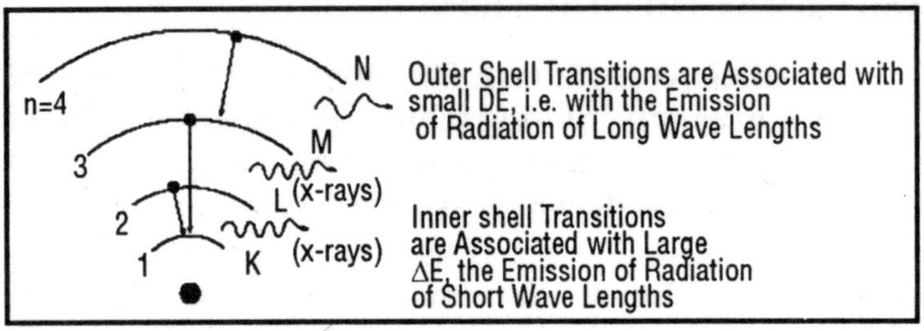

Fig. X-rays are Generated by Inner Shell Electron Transitions.

from n = 5 to n = 4 yield low energy radiation (long wavelength). For hydrogen, you recall, the wave number of the emitted radiation associated with a particular electron transition is given by the Rydberg equation:

$$\overline{V} = \left(\frac{1}{n_f^2} - \frac{1}{n_f^2}\right)R$$

For "hydrogen-like" atoms with the atomic number Z (containing one electron only) the corresponding Rydberg equation becomes:

$$\overline{V} = \left(\frac{1}{n_i^2} - \frac{1}{n_f^2}\right)RZ^2$$

From this relationship it is apparent that the energy difference associated with electron transitions increases strongly with the atomic number and that the wavelength of radiation emitted during such transitions moves with increasing Z from the 10^{-7} m range to the 10^{-10} m range (radiation now defined as X-rays).

To bring about such inner shell transitions requires the generation of an electron vacancy: an electron must be removed, for example from the K shell (n = 1), of an atom. Such a vacancy is conveniently produced in an X-ray tube by an electron beam (generated by a heated filament which is made a cathode) impinging, after being subjected to an accelerating potential of several kV, into a target material made anode. The impinging electrons will transfer part of their energy to electrons of the

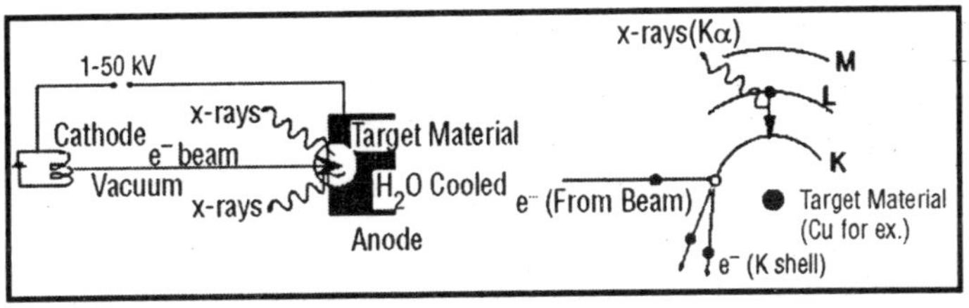

Fig. Generation of x-rays.

target material and result in electronic excitation. If the energy of the arriving electrons is high enough, some may knock out a K shell electron in the target and thus generate a vacancy. [It should be clear that a K → L excitation cannot take place since the L shell is filled: excitation must involve $(n = 1) \rightarrow (n = \infty)$.] When such a vacancy is generated, it can readily be filled by an electron from the L shell or the M shell of the same atom.

These internal electron transitions give rise to the emission of "characteristic" X-radiation which, because of its short wavelength, has extremely high "penetrating" power. [Since an electron beam is used to generate X-rays, the X-ray tube has to be evacuated: to dissipate the energy flux arriving at the target, the anode support (onto which the target is mounted) is water-cooled.]

Under standard operating conditions, the characteristic radiation emitted by the target comprises two sharp lines, referred to as Kα and Kβ lines. They are associated, respectively, with electron transitions from $n = 2$ to $n = 1$ and from $n = 3$ to $n = 1$.

Emitted X-ray spectra were extensively studied by H.G.J. Moseley who established the relationship between the wavelength of characteristic radiation and the atomic number Z of the radiation emitting target material. Experimentally he found that the Kα lines for various target materials (elements) exhibit the relationship:

$$\lambda K_{\alpha} \propto \frac{1}{Z^2} (\overline{V} K_{\alpha} \propto Z^2)$$

Moseley's empirical relationship (which reflects a behaviour in agreement with the Rydberg equation) can be quantified. While the energy levels associated with outer electron transitions are significantly affected by the "screening" effect of inner electrons (which is variable and cannot as yet be determined from first principles), the conditions associated with X-ray generation are simple. Very generally, the screening effect of the innermost electrons on the nuclear charge is accounted for in an effective nuclear

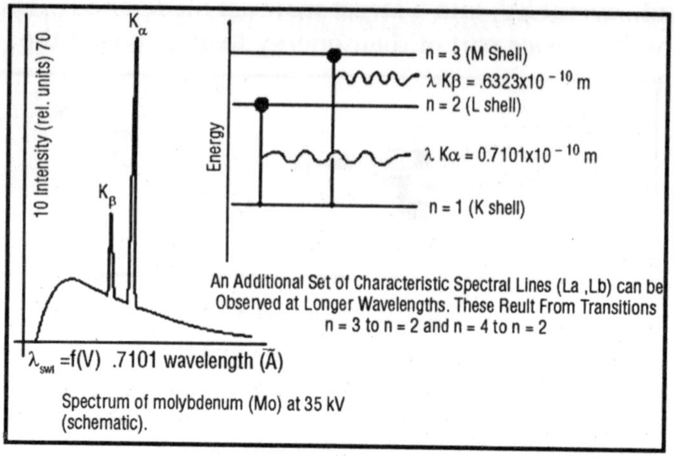

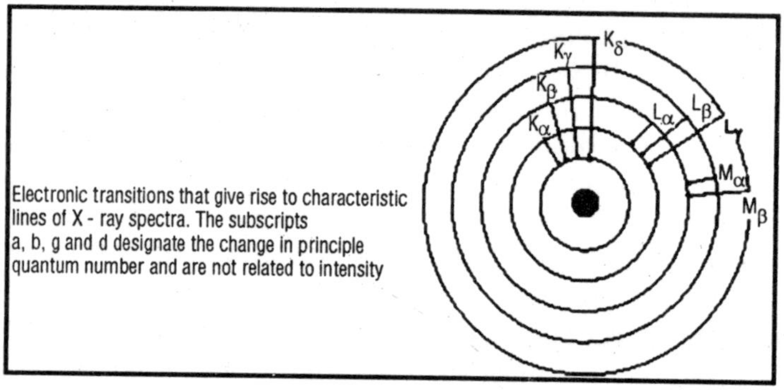

Fig. Electronic Transitions Giving Rise to Characteristic X- ray Spectra.

charge (Z–σ) and the Rydberg equation assumes the form:

$$\overline{V} = \left(\frac{1}{n_2^i} - \frac{1}{n_f^2} \right) R(z - \sigma)^2$$

where σ = screening effect

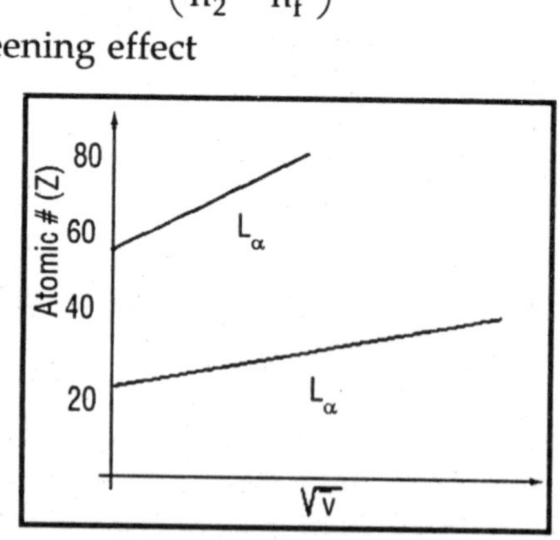

Fig. Moseley Relationship for Ka and La Radiation.

Considering the transition $n_2 \rightarrow n_1$, screening of the full nuclear charge is only provided by the one electron remaining in the K shell. Thus it is possible to use the modified Rydberg equation, taking σ = 1.Accordingly, we have:

$$\overline{V} = \left(\frac{1}{2^2} - \frac{1}{1^2}\right)R(z-1)^2 = -\frac{3}{4}R(Z-1)^2$$

where: R = Rydberg constant and Z = atomic number of the target material. [The minus sign (–) only reflects radiative energy given off by the system.] Similarly, for the characteristic La?series of spectral lines (n = 3 to n = 2) we find, after removal of one L electron, that the screening of the electrons in the K shell and the remaining electrons in the L shell reduces the nuclear charge by 7.4 (empirical value).

$$\overline{V}L_{\alpha} = \left(\frac{1}{3^2} - \frac{1}{2^2}\right)R(z-7.4)^2 = -\frac{5}{36}R(Z-7.4)^2$$

A second look at the X-ray spectrum of a Mo target, obtained with an electron accelerating potential of 35 kV, shows that the characteristic radiation appears superimposed on a continuous spectrum of lower and varying intensity. This continuous spectrum is referred to as bremsstrahlung (braking radiation) and has the following origin.

Electrons, impinging on the target material, may lose their energy by transferring it to orbiting electrons, as discussed above; in many instances, however, the electrons may come into the proximity of the force fields of target nuclei and, in doing so, will be "slowed down" or decelerated to a varying degree,

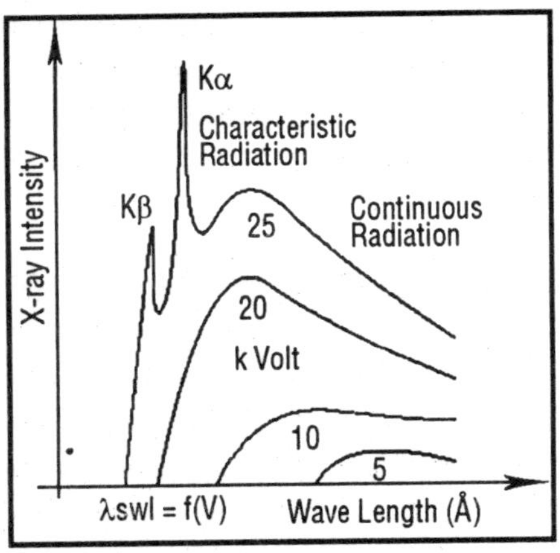

Fig. X-ray Spectrum of Mo Target as a Function of Applied Voltage.

ranging from imperceptible deceleration to total arrest. The energy lost in this slowing down process is emitted in the form of radiation (braking radiation, or bremsstrahlung). This energy conversion, as indicated, can range from partial to complete.

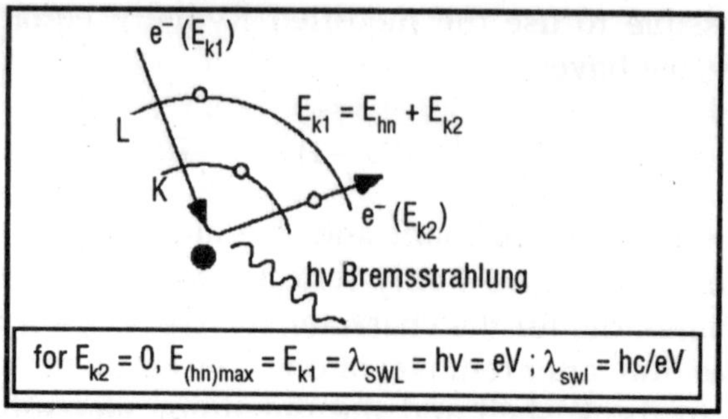

Fig. Origin of Bremsstrahlung; the Continuous Part of the X-ray Spectrum.

The incident electrons have an energy of e.V (electronic charge times accelerating Voltage) in the form of kinetic energy ($mv^2/2$), and their total energy conversion gives rise to a Shortest Wavelength (SWL) – the cut-off of the continuous spectrum for decreasing values. Analytically, we have:

$$eV = hv_{max} = h\frac{c}{\lambda_{SWL}}; \lambda_{SWL} = \frac{hc}{eV}$$

From this relationship it is evident that the cut-off of the continuous spectrum toward decreasing λ's (λ_{SWL}) is controlled by the accelerating potential.

THE "FINE STRUCTURE" OF CHARACTERISTIC X-RAYS

It is customary to consider the characteristic X-ray spectral lines as discrete lines. In reality, they are not discrete since the electron shells involved in the associated electron transitions have energy sublevels (s, p, d orbitals). These sublevels give rise to a "fine structure" insofar as the Kα lines are doublets composed of Kα and $K_{\alpha 2}$ lines. Similarly, Lα, Lβ etc., exhibit a fine structure.

These considerations suggest that X-ray spectra contain information concerning the energetics of electronic states. Obviously, analysis of X-rays emitted from a target of unknown composition can be used for a quantitative chemical analysis. [This approach is taken routinely in advanced scanning electron microscopy (SEM) where X-rays, generated by the focused electron-beam, are analysed in an appropriate spectrometer.] In fundamental studies it is also of interest to analyse soft X-ray spectra. For example, take the generation of X-rays in sodium (Na). By generating an electron vacancy in the K shell, a series of Kσand Kσlines will result. The cascading electron generates vacancies in the 2p level, which in turn can be filled by electrons entering from the 3s level (generation of "soft" X-rays).

If the X-rays are generated in a Na vapour, the 3s σ2p transition will yield a sharp line; on the other hand, if X-rays are analysed in sodium metal, the same transition results in the emission of a continuous broad band, about 30 Å in width. This finding confirms the existence of an energy band. An analysis of the width and intensity distribution of the X-ray band provides experimental data concerning the energy band width and the energy state density distribution within the energy band).

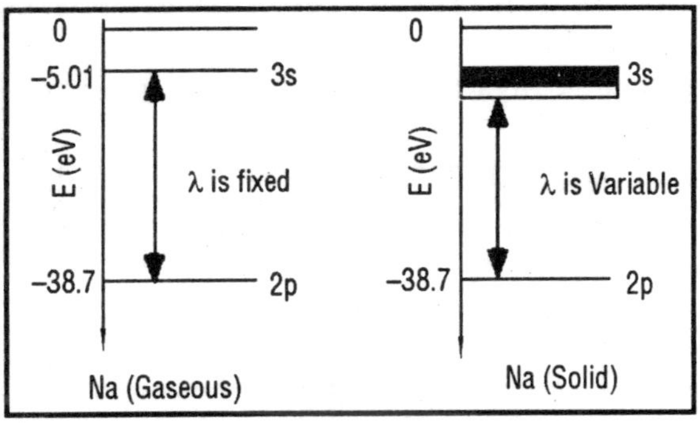

Fig. Soft X-rays from 3s - 2p Transitions in Solid Na and Na Vapour.

USE OF X-RAYS FOR STRUCTURAL ANALYSIS

The extensive use of X-rays for the analysis of atomic structural arrangements is based on the fact that waves undergo a phenomenon called diffraction when interacting with systems (diffracting centers) which are spaced at distances of the same order of magnitude as the wavelength of the particular radiation considered.

X-ray diffraction in crystalline solids takes place because the atomic spacings are in the 10^{-10} m range, as are the wavelengths of X-rays.

DIFFRACTION AND BRAGG'S LAW

The atomic structure of crystalline solids is commonly determined using one of several different X-ray diffraction techniques. Complementary structure information can also be obtained through electron and neutron diffraction. In all instances, the radiation used must have wavelengths in the range of 0.1 to 10 Å because the resolution (or smallest object separation distance) to which any radiation can yield useful information is about equal to the wavelength of the radiation, and the average distance between adjacent atoms in solids is about 10^{-10}m (1 Å).

Since there is no convenient way to focus X-rays with lenses and to magnify images, we do not attempt to look directly at atoms. Rather, we consider the interference effects of X-rays when scattered by the atoms, comprising a crystal lattice. This is analogous to studying the structure of an optical diffraction grating by examining the interference pattern produced when we shine visible light on the grating. (The spacing of lines on a grating is about 0.5 to 1 μm and the wavelength of visible radiation ranges from 0.4 to 0.8 m.) In the optical grating the ruled lines act as scattering centers, whereas in a crystal it is the atoms (more correctly, the electrons about the atom) which scatter the incident radiation. The geometrical conditions which must be satisfied for diffraction to occur in a crystal were first established by Bragg. He considered a monochromatic (single wavelength) beam of X-rays with coherent radiation (X-rays of common wave front) to be incident on a crystal.

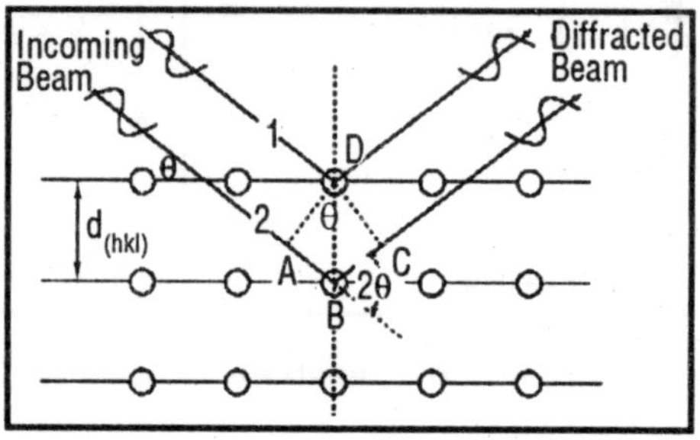

Fig. Bragg's Law, Assuming the Planes of Atoms Behave as Reflecting Planes.

Moreover, he established that the atoms which constitute the actual scattering centers can be represented by sets of parallel planes (in which the atoms are located) which act as mirrors and "reflect" the X-rays. In cubic systems the spacing of these planes, $d_{(hkl)}$, is related to the lattice constant

$$d_{(hkl)} = \frac{a}{\sqrt{h^2 + k^2 + l^2}}$$

For constructive interference of the scattered X-rays (the appearance of a diffraction peak) it is required that the beams, scattered on successive planes, be "in phase" (have again a common wave front) after they leave the surface of the crystal. In terms of the beams labeled 1 and 2 in figure this requires that the distance AB + BC be equal to an integral number of wavelengths of the indicent radiation. Accordingly:

$$\overline{AB} + \overline{BC} = n\lambda (n = 1,2,3,...)$$

$$\text{Since} \overline{AB} = \overline{BC} \text{and} \sin\theta = \frac{\overline{AB}}{d_{(hkl)}} \left[\overline{AB} = d_{(hkl)} \sin\theta\right]:$$

$$n\lambda = 2d_{(hkl)} \sin\theta$$

This relation is referred to as Bragg's Law and describes the angular position of the diffracted beam. In most instances of interest we deal with first order diffraction (n = 1) and, accordingly, Bragg's law is:

$$\lambda = 2d_{(hkl)} \sin\theta$$

[We are able to make n = 1 because we can always interpret a diffraction peak for n = 2, 3, ... as diffraction from (nh nk nl) planes – i.e., from planes with one-nth the interplanar spacing.]

If we consider figure as representative for a "diffractometer" set-up , we have a collimated beam of X-rays impinging on a (100) set of planes and at 2θ to the incident beam a detector which registers the intensity of radiation.

For a glancing incident beam (small θ) the detector will register only background radiation. As θ increases to a value for which 2d sin θ = λ, the detector will register high intensity radiation – we have a diffraction peak. From the above it is evident that the diffraction angle (θ) increases as the interplanar spacing, $d_{(hkl)}$, decreases.

The diffraction experiment as presently considered is intended to provide quantitative information on the volume (the lattice constant a) and shape characteristics (SC, BCC, FCC) of the unit cell. The intensity of diffraction peaks depends on the phase relationships between the radiation scattered by all the atoms in the unit cell. As a result, it happens quite often that the intensity of a particular peak, whose presence is predicted by Bragg's law, is zero. (This is because Bragg's law deals not with atom positions, but only with the size and shape of the unit cell.)

For example, consider the intensity of the (100) diffraction peak of a crystal which has a BCC unit cell. The phase relationships show that the X-rays scattered at the top and bottom faces of the unit cell, (100) planes, interfere constructively, but are 180° out of phase with the X-rays scattered by the atom at the centre of the unit cell. The resultant intensity is therefore zero. The rules which govern the presence of particular diffraction peaks in the different cubic Bravais lattices.

Table. Selection Rules for Diffraction Peaks in Cubic Systems

Bravais Lattice	**Reflections Present**	**Reflections Absent**	
Simple Cubic	All	None	
Body-Centered	Cubic	(h+k+l) = even	(h+k+l) = odd
Face-Centered	Cubic (either all odd or all even)	h,k,l unmixed	h,k,l mixed

The rules given are strictly true only for unit cells where a single atom is associated with each lattice point. (Unit cells with more than one atom per lattice point may have their atoms arranged in positions such that reflections cancel. For example, diamond has an FCC Bravais lattice with two atoms per lattice point.

All reflections present in diamond have unmixed indices, but reflections are missing. The fact that all reflections present have unmixed indices indicates that the Bravais lattice is FCC – the extra missing reflections give additional information as to the exact atom arrangement.)

A HYPOTHETICAL DIFFRACTION EXPERIMENT

A material is known to be of simple cubic structure; determine a, the lattice constant, by X-ray diffraction. In theory, the question may be answered by placing the crystal into a diffractometer, rotating it into all possible positions relative to the incident X-ray beam and recording all diffracting q values.

From the above we know that the smallest observed value must correspond to diffraction on {100} planes and also that d(100) = a. We may now use Bragg's equation to determine a, the lattice constant:

$$\lambda = 2d\sin\theta = 2a\sin\theta$$

$$a = \frac{\lambda}{2\sin\theta}$$

There are two simplifying assumptions in this problem:

- We know the system is SC.
- We are able, through rotation, to bring all planes present into diffraction conditions.

EXPERIMENTAL APPROACHES TO X-RAY DIFFRACTION

In the context of this course we are interested in making use of X-ray diffraction for the purpose of (a) identifying (cubic) crystal systems, (b) determining the lattice constant, a, and (c) identifying particular planes or meaningful orientations.

The possible approaches can, in principle, be identified through an examination of Bragg's law. The Bragg condition for particular d(hkl) values can be satisfied by adjusting either one of two experimental variables: (a) λ, the wavelength of the X-ray beam used, or (b) θ, the orientation of the crystal planes relative to the incident X-rays.

(a) Fixed θ, Variable λ: One means of satisfying Bragg's law is to irradiate a stationary single crystal (θ fixed for all planes within the crystal) with an X-ray beam of "white" radiation, which contains the characteristic and continuous spectrum produced by an X-ray tube. (For λ variable we

have the simultaneous exposure of a crystal to a range of λvalues). Each set of planes will reflect (diffract) the particular λwhich satisfies the Bragg condition for the fixed θ. The diffracted beams may conveniently be recorded with a Polaroid camera or, alternately, with an electronic imaging device. It is possible to analyse either the transmitted or the back-reflected X-rays.

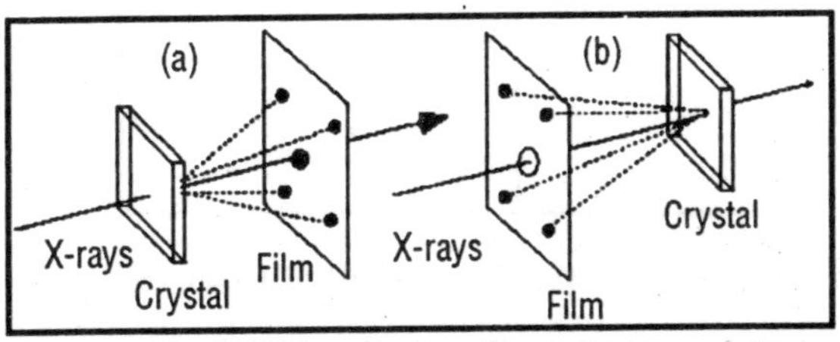

Fig. Laue Diffraction in (a) Transmission and (b) Back-reflection Mode.

This experimental procedure is referred to as the Laue technique; it is mostly conducted in the back-reflection mode. Note that the approach taken makes it possible to determine the values of sfor each reflection, but not the corresponding l.

Therefore, the technique cannot be used, for example, to determine lattice constants. However, it is very valuable if particular planes or crystal orientations are to be identified.

(b) *Fixed λ(Monochromatic X-Rays), Variable* θ: The basic prerequisite for this approach is the availability of a monochromatic X-radiation of known wavelength (λ). Such radiation can be conveniently obtained by using a crystal (i.e., its diffracting property) as a filter or monochromator. Filter action is achieved by positioning the crystal in such a way that the unfiltered radiation emitted by the X-ray tube becomes incident at an angle, θ, on a set of low index planes which satisfy Bragg's law for the highest intensity radiation emitted.

The condition of a fixed lλ and variable l θ is experimentally used in two techniques. Using a diffractometer, we place a sample (ground to a powder) into the centre of a rotating stage and expose it to a monochromatic X-ray beam. The sample is rotated into diffraction condition and the diffraction angle determined.

In the Debye-Scherrer method the sample is ground to a powder and placed (in an ampoule) into the centre of a Debye-Scherrer camera. Exposed to monochromatic X-rays, in this way a large number of diffracted cone-shaped beams are generated such that the semiangles of the cones measure 2θ, or twice the Bragg angle for the particular diffracting crystallographic planes.

The reason diffracted beams are cone-shaped is that the planes in question (within the multitude of randomly oriented grains) give rise to diffraction for any orientation around the incident beam as long as the incident beam forms the appropriate Bragg angle with these planes – thus there is a rotational symmetry of the diffracted beams about the

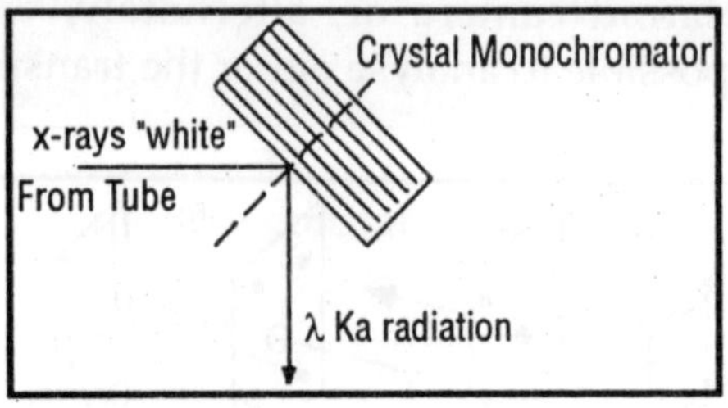

Fig. Isolation of Monochromatic Radiation from Target Radiation.

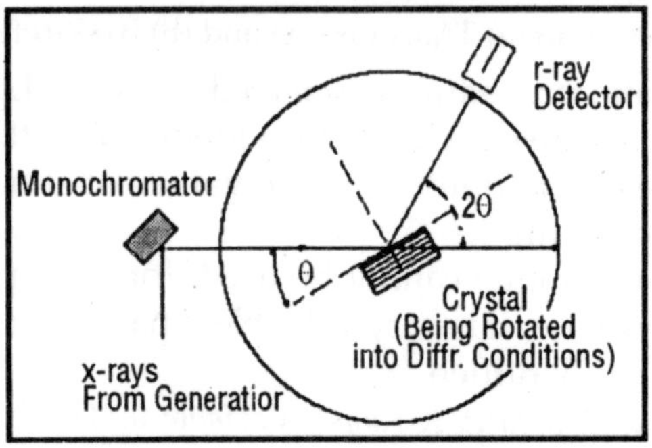

Fig. X-ray Diffractometer Setup.

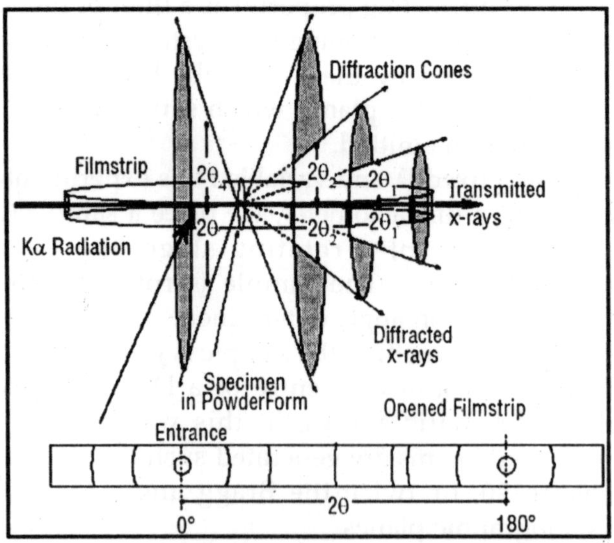

Fig. Debye-Scherrer Powder Diffraction Setup and Analysis.

direction of the incident beam. Those planes with the largest interplanar spacing have the smallest Bragg angle, θ.

In a Debye–Scherrer arrangement, after exposing a powder of a crystalline material to monochromatic X-rays, the developed film strip will exhibit diffraction patterns such. Each diffraction peak (dark line) on the film strip corresponds to constructive interference at planes of a particular interplanar spacing. The problem now consists of "indexing" the individual lines – i.e., determining the Miller indices for the diffraction lines:

$$\text{Bragg}: \lambda = 2d_{(hkl)} \sin\theta; d_{(hkl)} = \frac{a}{\sqrt{h^2 + k^2 + l^2}}$$

$$\lambda^2 = 4d_{(hkl)}^2 \sin^2\theta; d_{(hkl)}^2 = \frac{a^2}{(h^2 + k^2 + l^2)}$$

Substitution and rearrangement of above yields:

$$\frac{\sin^2\theta_1}{(h^2 + k^2 + l^2)_1} = \frac{\sin^2\theta_2}{(h^2 + k^2 + l^2)_2} = \frac{\sin^2\theta_3}{(h^2 + k^2 + l^2)_3} = \text{const.}$$

holds. Since the sum $(h^2 + k^2 + l^2)$ is always integral and $\lambda^2/4a^2$ is a constant, the problem of indexing the pattern of a cubic system is one of finding a set of integers $(h^2 + k^2 + l^2)$ which will yield a onstant quotient when divided one by one into the observed sin2σ values. (Certain integers such as 7, 15, 23, etc. are impossible because they cannot be formed by the sum of three squared integers.)

Indexing in step-by-step sequence is thus performed as follows: θvalues of the lines are obtained from the geometric relationship of the unrolled film strip. Between the exit hole of the X-ray beam and the entrance hole (2θ= 180°) the angular relationship is linear. The increasing qvalues for successive lines are indexed θ_1, θ_2,$\theta_3$etc., and sin2θis determined for each.

If the system is simple cubic we know that all planes present will lead to diffraction and the successive lines result from diffraction on planes with decreasing interplanar spacing: (100), (110), (111), (200), (210), (211), (220), etc. From equation above we recognize:

$$\frac{\sin^2\theta_1}{1} = \frac{\sin^2\theta_2}{2} = \frac{\sin^2\theta_3}{3} = \frac{\sin^2\theta_4}{4} = \frac{\sin^2\theta_5}{5} = \text{const.}$$

If the system is BCC, however, we know from the selection rules that only planes for which (h + k + l) = even will reflect. Thus:

$$\frac{\sin^2\theta_1}{3} = \frac{\sin^2\theta_2}{4} = \frac{\sin^2\theta_3}{6} = \frac{\sin^2\theta_4}{8} \text{etc.} = \text{const.}$$

[SC can be differentiated from BCC through the fact that no sum of three squared integers can yield 7, but 14 can be obtained from planes].

For FCC systems, the selection rules indicate reflections on planes with unmixed h,k,l indices:

$$\frac{\sin^2\theta_1}{3}=\frac{\sin^2\theta_2}{4}=\frac{\sin^2\theta_3}{8}==\text{const.}$$

After proper indexing, the constant is obtained:

$$\frac{\sin^2\theta}{(h^2+k^2+l^2)}=\text{const.}$$

and the particular Bravais lattice is identified. The lattice constant of the unit cell is subsequently obtained, knowing the wavelength of the incident radiation:

$$\frac{\sin^2\theta}{(h^2+k^2+l^2)}=\text{const.}=\frac{\lambda_2}{4a^2}$$

$$a^2=\frac{\lambda^2}{4\sin^2\theta}(h^2+k^2+l^2)$$

$$a=\frac{\lambda}{2\sin\theta}\sqrt{(h^2+k^2+l^2)}$$

Chapter 3

Imperfect Solid State

Real crystals are never perfect: they always contain a considerable density of defects and imperfections that affect their physical, chemical, mechanical and electronic properties. The existence of defects also plays an important role in various technological processes and phenomena such as annealing, precipitation, diffusion, sintering, oxidation and others.

It should be noted that defects do not necessarily have adverse effects on the properties of materials. There are many situations in which a judicious control of the types and amounts of imperfections can bring about specific characteristics desired in a system. This can be achieved by proper processing techniques. In fact, "defect engineering" is emerging as an important activity.

All defects and imperfections can be conveniently considered under four main divisions: point defects, line defects or dislocations, planar defects or interfacial or grain boundary defects, and volume defects. We can also add here macroscopic or bulk defects such as pores, cracks and foreign inclusions that are introduced during production and processing of the solid state.

Point defects are inherent to the equilibrium state and thus determined by temperature, pressure and composition of a given system. The presence and concentration of other defects, however, depend on the way the solid was originally formed and subsequently processed.

Briefly consider the effects of imperfections or crystal defects on a few important properties of solids. The electrical behaviour of semiconductors, for example, is largely controlled by crystal imperfections. The conductivity of silicon can thus be altered in type (n or p) and by over eight orders of magnitude through the addition of minute amounts of electrically active dopant elements. In this case, each atom of dopant, substitutionally incorporated, represents a point defect in the silicon lattice.

The fact that such small amounts of impurity atoms can significantly alter the electrical properties of semiconductors is responsible for the development of the transistor and has opened up the entire field of solid state device technology. Practically none of the semiconducting properties that led to these engineering accomplishments are found in a "perfect" crystal. They are properties peculiar to the defective solid state. The existence of dislocations (line defects) in crystals provides a mechanism by which permanent change of shape or mechanical deformation can occur.

A crystalline solid free of dislocations is brittle and practically useless as an engineering material. While the existence of dislocations in crystals insures ductility (ability to deform), the theoretical strength of crystalline solids is drastically reduced by their presence.

We should recognize that dislocations play a central role in the determination of such important properties as strength and ductility. In fact, virtually all mechanical properties of crystalline solids are to a significant extent controlled by the behaviour of line imperfections.

The ability of a ferromagnetic material (such as iron, nickel or iron oxide) to be magnetized and demagnetized depends in large part on the presence of two-dimensional imperfections known as Bloch walls. These interfaces are boundaries between two regions of the crystal which have a different magnetic state.

As magnetization occurs, these defects migrate and by their motion provide the material with a net magnetic moment. Without the existence of Bloch walls all ferromagnetic materials would be permanent magnets. In fact, electromagnets would not exist if it were not for this type of defect.

The presence of surface defects such as cracks causes brittle materials like glass to break at small applied stresses. This fact is familiar to anyone who has broken a glass tube by first filing a small notch (or crack) into the surface.

Removal of cracks from the surface of glass either by etching in hydrofluoric acid or by flame polishing almost always raises the fracture strength. For example, glass in the absence of any surface cracks has a fracture strength of $\sim 10^{10}$ Newton/m^2 (as opposed to real glass which has a fracture strength of $\sim 10^7$ Newton/m^2).

POINT DEFECTS

Formation of Point Defects

An incontrovertible law of nature states: "Nothing is perfect". This law applies to humans as well as to the inorganic world of crystalline solids and can be formulated as the 2nd law of thermodynamics:

$$F = H - TS$$

where F is the free energy of a given system, H is the heat content or enthalpy and TS is the entropy, or disorder, term. If a reaction takes place at a temperature T, we find the change in F (Δ F) related to a change in H (Δ H), the heat content, and possibly also a change in TS (TΔ S). Such is the case when defects are formed in a perfect solid: The energy distribution in a solid (Maxwell-Boltzmann) suggests that a number of individual atoms may acquire enough thermal energy to be displaced from the equilibrium lattice site into an interstitial position.

This process of point defect formation requires energy and leads to lattice strain which constitutes, as discussed earlier, an increase in the heat content of the system (Δ H is positive and increases linearly with the number of defects formed). The departure from perfection by the generation of defects leads to disorder (Δ S is positive). The magnitude of disorder generated (Δ S) is very large during the initial step from perfection to slight disarray, but the increase in disorder (with a given number of defects generated) decreases as the overall disorder increases.

Correspondingly the term TΔ S drops rapidly at the beginning and then flattens out. The net result, free energy, exhibits a minimum for a certain number of defects in the solid [equilibrium defect density = f (temperature)]; the Fminimum suggests also that the transition from perfection to equilibrium defect structure is spontaneous: it occurs naturally.

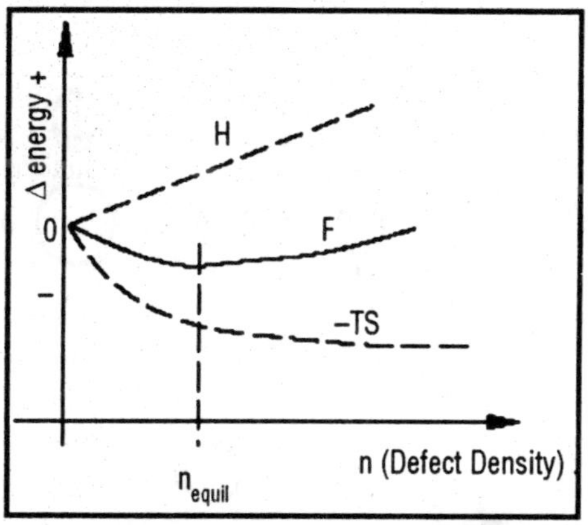

Fig. Thermodynamics of Point Defect Formation.

While the detailed mechanisms for the formation of atomic vacancies in solids are still the subject of extensive research, the associated equilibrium energetics are clear: calculations of the thermal energy of atoms in a lattice show that the average vibrational energy of lattice atoms is much less than 1 eV (the approximate energy change associated with vacancy formation, i.e., the least amount of energy required to form a vacancy) at room temperature.

Therefore a lattice atom will only acquire the energy ΔH_d, the energy required to form the defect, upon the occurrence of a large energy fluctuation. Since the relative probability of an atom having an energy ΔH_d or more in excess of the ground state energy is $e^{-\Delta H_d/kT}$,, the probability that an atomic site is vacant varies in the same way. In a (molar) crystal containing N atomic sites, the number n_d of vacant sites is, therefore,

$$n_d = ANe^{-\Delta H_d/kT}$$

n_d = The number of defects (in equilibrium at T)

N =The total number of atomic sites per mole

ΔH_d =The energy necessary to form the defect

T = The absolute temperature (K)

k = The Boltzmann constant

A = proportionality constant

Point Defects in "Pure" Metallic Systems

Point defects in "pure" crystalline metals are defects of atomic dimensions, such as impurity atoms, the absence of a matrix atom and/or the presence of a matrix atom in the wrong place. Some of these point defects. An impurity atom that

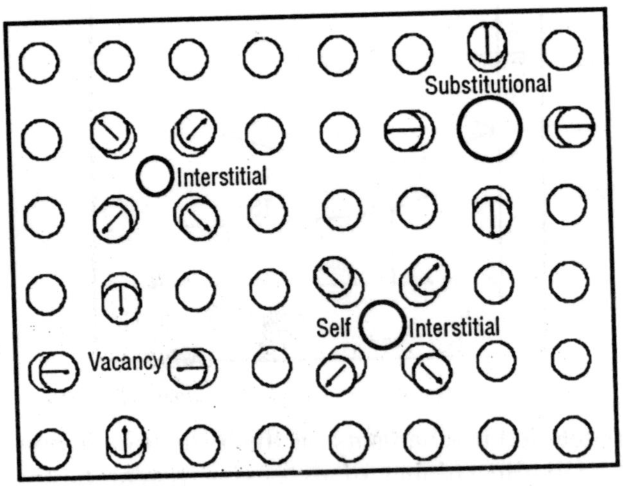

Fig. Point Defects in Crystalline Solids.

or interstitial site depends largely on the size of the atom relative to the size of the site. Small atoms are usually interstitial impurities, while larger atoms are usually substitutional impurities.

A vacancy is an atom site, normally occupied in the perfect crystal, from which an atom is missing. Often the term "vacancy" is used to denote a so-called Schottky defect, which is formed when an atom or an ion leaves a normal lattice site and repositions itself in a lattice site on the surface of the crystal.

This may be the result of atomic rearrangement in an existing crystal at a high temperature when atomic mobility is high because of increased thermal vibrations.

A vacancy may also originate in the process of crystallization as a result of local disturbances during the growth of new atomic planes on the crystal surface. Vacancies are point defects of a size nearly equal to the size of the original (occupied) site; the energy of the formation of a vacancy is relatively low - usually less than 1 eV.

The number of vacancies at equilibrium at each temperature in a crystal can be determined from equation, in which ΔH_d is the energy necessary to take an atom from a regular site of the crystal and place it on the surface for a Schottky-type defect. When a solid is heated a new higher equilibrium concentration of vacancies is established, usually first at crystal surfaces and then in the vicinity of dislocations and grain boundaries which provide sites for the atoms which have left their normal lattice site.

Vacancies gradually spread throughout the crystal (from the surfaces into the bulk). On cooling the vacancy concentration is lowered by "diffusion of vacancies" to grain boundaries or dislocations, which act as sinks. In both cases, the new equilibrium vacancy concentration is established only after a finite amount of time. The rate at which vacancies move from point to point in the lattice decreases exponentially with decreasing temperature.

Thus, on very rapid cooling (quenching) from a high temperature near the melting point most of the vacancies do not have time to diffuse to sinks and are said to be "frozen in". This gives a considerably greater ("non-equilibrium") concentration of vacancies in quenched specimens than that indicated by the thermal equilibrium value.

The concentration of vacant lattice sites in pure materials is very small at low temperatures - about one vacancy every 108 atom sites - and increases with increasing temperature to about one vacancy every 103 sites at the melting temperature. Vacancies are important because they control the rate of matrix (or substitutional) atom diffusion - i.e., atoms are able to move around in a crystalline solid primarily because of the presence of vacancies.

The mechanism by which they move is the same as that associated with moving a car in a filled parking lot to the exit. Self-interstitials are generally not encountered in close-packed

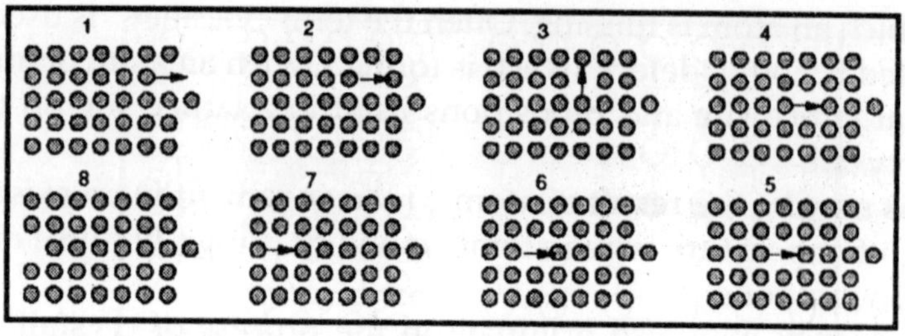

Fig. Dynamics of Vacancy Movements in a Close Packed Solid.

metallic systems, but may be introduced by irradiation. For example, high-energy neutrons from atomic fission can knock metal atoms from their regular sites into interstitial sites, creating vacancy-interstitial pairs.

Point Defects in Ionic Solids

Point defects in ionic structures differ from those found in pure elements because of the charge neutrality requirement. For example, in a pure monovalent ionic material a cation vacancy must have associated with it either a cation interstitial or an anion vacancy to maintain charge neutrality.

Similar requirements hold for anion vacancies. A vacancy pair defect (migration of a cation and an anion to the surface) is usually called a Schottky imperfection, and a vacancy-interstitial pair defect is referred to as a Frenkel imperfection (an anion or cation has left its lattice position, which becomes a vacancy, and has moved to an interstitial position).

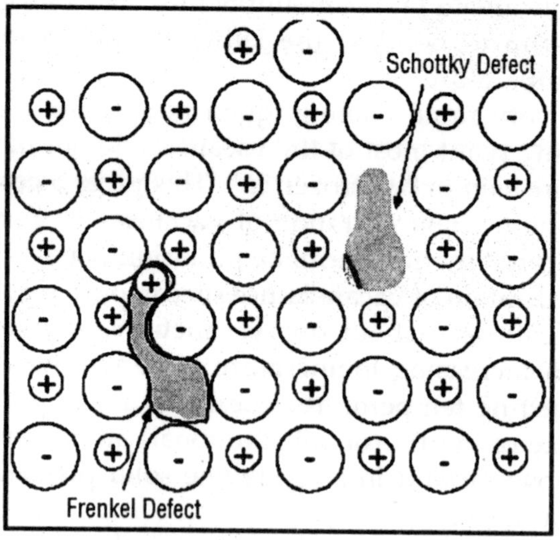

Fig 4 Point Defects in Ionic Solids.

Self-interstitials are much more common in ionic structures than in pure elements because many ionic compounds have relatively large interstitial sites available. That is, there are often interstitial sites in the unit cell that have nearly the same surroundings as normal atom sites. (For example, in BeO the Be atoms fill only one-half the available tetrahedral sites, leaving four possible cation interstitial sites per unit cell. Thus a Be atom could go from a regular lattice site to an almost equivalent interstitial site with little distortion of the lattice.)

Foreign atoms in ionic crystals produce defects that also must maintain charge neutrality. For example, in NaCl a monovalent cation, such as lithium, may simply replace one of the sodium ions as a substitutional impurity. But a divalent cation, such as calcium, replacing a sodium ion must be accompanied by either a cation vacancy or an anion interstitial if charge neutrality is to be maintained.

Correspondingly, monovalent impurity cations in a divalent structure (e.g., Na in MgO) must be accompanied by an appropriate number of cation interstitials or anion vacancies.

Point Defects in Covalently Bonded Solids

Substitutional impurities in covalently bonded materials can create a unique imperfection in the electronic structure if the impurity atom is from a group in the periodic table other than the matrix atoms. For example, you already considered Group V and Group III elements in a Group IV matrix, such as As or B in Si.

When foreign atoms are incorporated into a crystal structure, whether in substitutional or interstitial sites, we say that the resulting phase is a solid solution of the matrix material (solvent) and the foreign atoms (solute). The term "solid solution", however, is not restricted to the low solute contents of doped semiconductor systems; there are many solid solutions, such as metallic alloys, that comprise a wide composition range.

LINE DEFECTS

Line imperfections, or dislocations, in crystalline solids are defects that produce lattice distortions centered about a line. A dislocation is simply the edge of an extra inserted fractional plane of atoms. Normally the symbol ⊥ is used to represent a positive

dislocation (extra fractional plane) is used to represent a negative dislocation (missing fractional plane).

The importance of dislocations is readily demonstrated in the deformation of crystalline materials. The plane in which a dislocation moves through the lattice is called a slip plane. With an applied shear stress the dislocation moves, atomic row by atomic row, and one part of the crystal is displaced relative to the other.

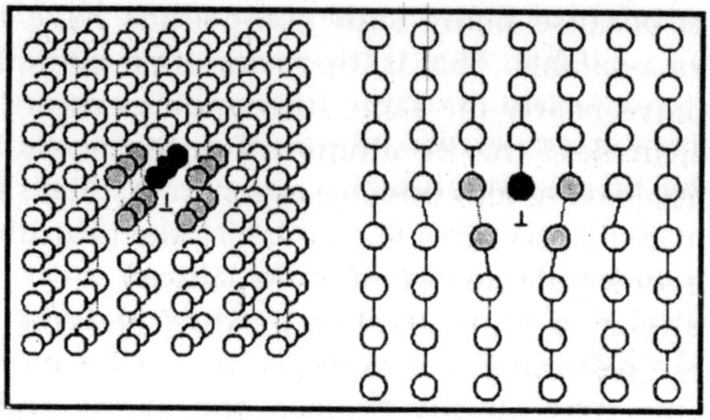

Fig. Schematic Presentation of a Dislocation; The Last Row of Atoms (Dark) in the Inserted Fractional Plane.

When the dislocation has passed through the crystal, the portion of the crystal above the slip plane has shifted one atomic distance relative to the portion below the slip plane. In other words, the motion of the dislocation has caused the crystal to change its shape - to be permanently deformed.

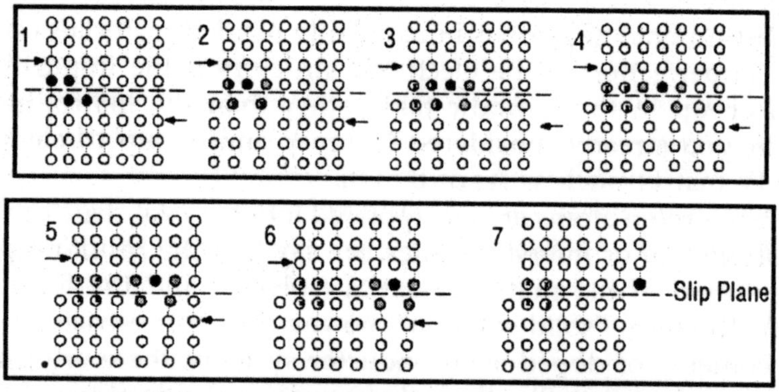

Fig. Plastic Deformation of Crystalline Solid by Slip Associated with Stress Induced Motion of Dislocation.

Please note: on either side of the dislocation the crystal lattice is essentially perfect, but in the immediate vicinity of the dislocation the lattice is severely distorted.

For a positive edge dislocation, the presence of the extra half plane causes the atoms above the slip plane to be put in compression, while those below the slip plane are put in tension. Consequently, the edge dislocation will have a stress field around it that is compressive above the slip plane and tensile below the slip plane.

Plastic Deformation By Slip

When single crystals of metal (or semiconductor) are pulled in tension, they will begin to deform (elongate) plastically at relatively low stress levels, and "blocks" of the crystals slide over one another because of dislocation motion. Simultaneously, so-called slip lines appear on their surface. It is found that deformation by slip occurs most easily on planes with high atomic density and with large interplanar spacing, while the direction of slip is in all instances an atomically "close-packed direction".

For FCC structures we therefore observe as the primary slip system {111} planes in <110> direction, while in BCC structures the primary slip occurs on {110} planes in <111> directions. (It should be noted that an alternate deformation mechanism is "deformation twinning", presently not to be considered.)

Dislocation Climb

Climb is the name given to the motion of dislocations when the extra "half" plane is extended farther into a crystal or partially withdrawn from it. Clearly, the climb process is not a motion of the plane, but rather its growth or shrinking as a result of the addition of atoms or "vacancies" respectively from the environment of the dislocation.

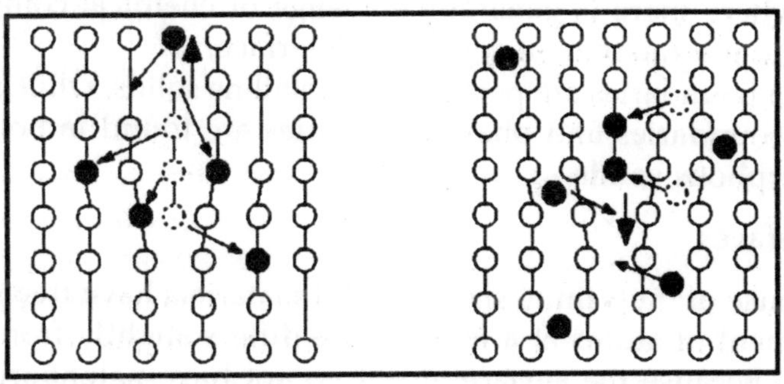

Fig. Dislocation Climb by (a) Loss of Atoms to Surrounding Vacancies and (b) Incorporation of Interstitial Atoms.

Multiplication of Dislocations

Since during slip each dislocation leaves the matrix, macroscopic deformation could not take place given normal dislocation densities in the range of 10^6-10^8/cm^3. Examination of the deformed crystals indicates that multiplication of dislocations takes place during deformation.

Dislocation Interactions

The relative ease with which dislocations move across a solid matrix can be attributed to the severe displacements of atoms in the core of dislocations. If these local stresses are reduced, the mobility of dislocations - and thus the ease of slip - is reduced.

It is found that impurities in the vicinity of dislocation cores tend to reduce the local distortion energy of the dislocations and thus stabilize the system against slip. In many systems impurities are intentionally added (e.g., solid solution hardening) to increase the strength of materials. Similarly, micro-precipitates tend to impede dislocation motion (e.g., precipitation hardening).

INTERFACIAL IMPERFECTIONS

The several different types of interfacial, or planar imperfections, in solids can be grouped into the following categories:

- Interfaces between solids and gases, which are called free surfaces;
- Interfaces between regions where there is a change in the electronic structure, but no change in the periodicity of atom arrangement, known as domain boundaries;
- Interfaces between two crystals or grains of the same phase where there is an orientation difference in the atom arrangement across the interface; these interfaces are called grain boundaries;
- Interfaces between different phases, called phase boundaries, where there is generally a change of chemical composition and atom arrangement across the interface.

Grain boundaries are peculiar to crystalline solids, while free surfaces, domain boundaries and phase boundaries are found in both crystalline and amorphous solids.

Free Surfaces

Because of their finite size, all solid materials have free surfaces. The arrangement of atoms at a free surface differs slightly from the interior structure because the surface atoms do not have neighboring atoms on one side. Usually the atoms near the surface have the same crystal structure but a slightly larger lattice parameter than the interior atoms.

Perhaps the most important aspect of free surfaces is the surface energy associated with surfaces of any solid. The source of this surface energy may be seen by considering the surroundings of atoms on the surface and in the interior of a solid. To bring an atom from the interior to the surface, we must either break or distort some bonds - thereby increasing the energy.

The surface energy is defined as the increase in energy per unit area of new surface formed. In crystalline solids, the surface energy depends on the crystallographic orientation of the surface - those surfaces that are planes of densest atomic packing are also the planes of lowest surface energy.

This is because atoms on these surfaces have fewer of their bonds broken or, equivalently, have a larger number of nearest neighbors within the plane of the surface. Typical values of surface energies of solids range from about 10^{-1} to 1 J/m^2. Generally, the stronger the bonding in the crystal, the higher the surface energy.

Surface energies can be reduced by the adsorption of foreign atoms or molecules from the surrounding atmosphere. For example, in mica the surface energy of freshly cleaved material in a vacuum is much higher than the surface energy of the same surface cleaved in air. In this instance, oxygen is adsorbed from the air to partially satisfy the broken bonds at the surface.

Impurity atom adsorption makes it almost impossible to maintain atomically clean surfaces. As a result, surface properties such as electron emission, rates of evaporation and rates of chemical reactions are extremely dependent on the presence of any adsorbed impurities. These properties will be different if the measurements are made under conditions giving different surface adsorption.

Grain Boundaries

Grain boundaries separate regions of different crystallographic orientation. The simplest form of a grain boundary is an interface composed of a parallel array of edge dislocations. This particular type of boundary is called a tilt boundary because the misorientation is in the form of a simple tilt about an axis, parallel to the dislocations. Tilt boundaries are referred to as low-angle boundaries because the angle of misorientation is generally less than 10°.

When a grain boundary has a misorientation greater than 10 or 15, it is no longer practical to think of the boundary as being made up of dislocations because the spacing of the dislocations would be so small that they would lose their individual identity. The grain boundary represents a region a few atomic diameters wide where there is a transition in atomic periodicity between adjacent crystals or grains.

Grain boundaries have an interfacial energy because of the disruption in atomic periodicity in the vicinity of the boundary and the broken bonds that exist across the interface. The interfacial energy of grain boundaries is generally less than that of a free surface because the atoms in a grain boundary are surrounded on all sides by other atoms and have only a few broken or distorted bonds.

Solids with grain boundaries are referred to as polycrystalline, since the structure is composed of many crystals - each with a different crystallographic orientation. In the case of iron the grain boundary structure can be revealed by preferential chemical attack (etching) at the grain boundaries, while the grain structure in polyethylene is revealed by the use of polarized light. The grain structure is usually specified by giving average grain diameter or by using a scheme developed by the American Society for Testing and Materials (ASTM). In the ASTM procedure the grain size is specified by a "grain size number" (n) where

$$N = 2^{n-1}$$

with N equal to the number of grains per square inch when the sample is viewed at 100X magnification. For example, at a magnification of X = 100, a material with grain size number 8 will show 128 grains per inch2 - this material in effect has (at X = 1) 1.28×10^6 grains per square inch. If the grains are approximately square in cross section, this corresponds to an average grain dimension of 8.8×10^{-4} in.

In polycrystalline samples the individual grains usually have a random crystallographic orientation with respect to one another, and the grain structure is referred to as randomly oriented. In some instances, however, the grains all have the same orientation to within a few degrees. In this instance the material is said to have a preferred orientation or texture.

Phase Boundaries

A phase is defined as a homogeneous, physically distinct and mechanically separable portion of the material with a given chemical composition and structure. Phases may be substitutional or interstitial solid solutions, ordered alloys or compounds, amorphous substances or even pure elements; a crystalline phase in the solid state may be either polycrystalline or exist as a single crystal.

Solids composed of more than one element may - and often do - consist of a number of phases. For example, a dentist's drill, something painfully familiar to all of us, consists of a mixture of small single crystals of tungsten carbide surrounded by a matrix of cobalt. Here the cobalt forms a continuous phase. Polyphase materials such as the dentist's drill are generally referred to as composite materials. Composite materials have great importance in the engineering world because they have many attractive properties that set them apart from single-phase materials.

For example, the dentist's drill has good abrasive characteristics (due to the hard carbide particles) and good toughness and impact resistance (due to the continuous cobalt matrix). Neither the tungsten carbide nor the cobalt has both abrasion resistance and impact resistance, yet the proper combination of the two phases yields a composite structure with the desired properties.

The nature of the interface separating various phases is very much like a grain boundary. Boundaries between two phases of different chemical composition and different crystal structure are similar to grain boundaries, while boundaries between different phases with similar crystal structures and crystallographic orientations may be analogous to low-angle grain boundaries in both energy and structure.

The concept of a solid consisting of a continuous phase and a discontinuous phase (or phases) leads to a simple classification of the various types of composite materials. Table gives this classification, which is based on the structure (whether amorphous or crystalline) of the continuous and discontinuous phases.

Table. Classification of Composite or Multiphase Materials.

Discontinuous Phase	Continuous Phase (or Phases)	Examples
Crystalline	Crystalline	All metallic systems such as cast iron, steel, soft solder, etc.; most natural rocks such as granite and marble.
Crystalline	Amorphous	None of practical Amorphous Crystalline
Amorphous	Crystalline	Most man-made ceramics such as building bricks and electrical insulator porcelain, concrete, partially crystalline polymers, some polymer-crystalline particle composites.significance.
Amorphous	Amorphous	Fiberglass, asphalt, wood, hydrated cement, other gels.

Chapter 4

Glasses

When a liquid phase is cooled to below its freezing temperature, it usually transforms into a crystalline solid, i.e. it crystallizes. Some liquids, because of complex molecular configuration or slow molecular transport, do not "crystallize" (assume an ordered configuration) upon being cooled to low temperatures, but instead form a rigid disordered network, known as glass, which is very similar in structure to that of the liquid.

Most metals and alkali halides crystallize easily upon cooling through the freezing temperature because the structural rearrangement from the liquid to the crystalline state is simple and bonding is nondirectional. At temperatures just above the freezing temperature, most metals and molten salts have fluidities approximately equivalent to that of water at room temperature.

Thus, the required structural rearrangements can take place relatively easily. In contrast, many inorganic silicates form glasses upon cooling because the fluidity at and even above the freezing temperature is very low. This glass formation is related to the high silicon-oxygen single-bond energies and the directional bonding requirements imposed by sp^3 hybridization of silicon.

The disordered liquid cannot flow easily and thus cannot undergo the rearrangements required for crystallization. Moreover, the per cent solubility in each other.

If a melt containing 50 wt.% silver and 50 wt.% copper is slowly cooled, two crystalline phases are produced. When this same alloy is obtained by vacuum deposition, a metastable metallic glass is formed. Initially the glass has metallic electrical conductivity comparable to that of the liquid alloy. When the glass is heated for 1/2 hour at 380K, it converts to a crystalline, metastable face-centered cubic solid solution.

Only on heating at about 500K does the equilibrium, two-phase structure form. Many pure nonmetallic elements, however, such as S, Se, Ge and Si, form a stable glass under suitable conditions.

THE GLASSY STATE

The properties and chemical composition of the glass used for optical purposes differ greatly from those of the glass used for baking dishes. Likewise, the glass used for window panes differs from the glass used for artwork or optical purposes.

However, (most) inorganic glasses have one thing in common: they are ceramic materials (compounds that contain metallic and non-metallic elements, most often oxygen) with the amorphous structure of a frozen liquid.

Quartz glass is not crystalline at room temperature because the rearrangement of interatomic bonds between Si and O required for the establishment of order is impeded: the viscosity of glass is so high at room temperature that millions of years would be required for it to crystallize.

Hence, glass can be viewed as a perfectly rigid "liquid", being exceptional in that its extraordinarily high viscosity slows down its flow properties even on an expanded time scale.

Upon cooling of liquid "glass" we observe that $\Delta V/\Delta T$ (the contraction with lowering of T) changes. This may be seen by following the three basic steps in the production of glass:

(1) The melting of quartz sand (minute crystals of silica),

(2) The shaping of the glass while in a viscous state

(3) The controlled cooling of the shaped article.

The SiO_4 tetrahedra in the sand (crystals) are arranged in an ordered pattern, but in

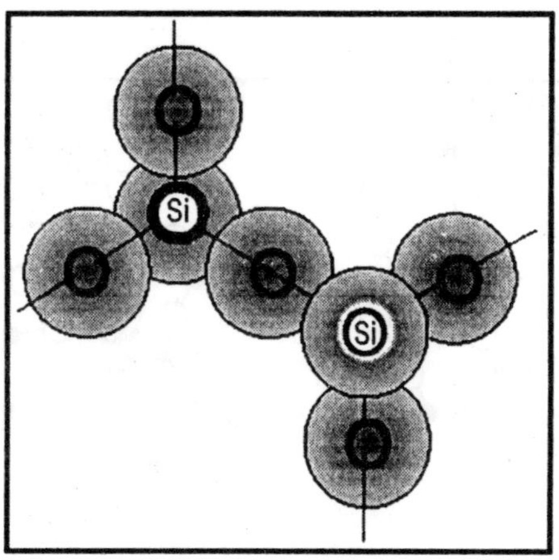

Fig. Basic Si - O - Si - Units as Established with sp3 Hybridization of Si.

the molten state the crystal matrix breaks up into strings and rings of tetrahedra in irregular patterns. Because the temperature is very high, the groupings continually break up and reform, making the mass fluid. Upon cooling of molten SiO_2 the tetrahedra form larger groupings and the glass becomes more viscous, permitting it to be shaped by blowing, rolling, pressing, etc. With continuing cooling, larger groupings of tetrahedra form as their kinetic energy decreases.

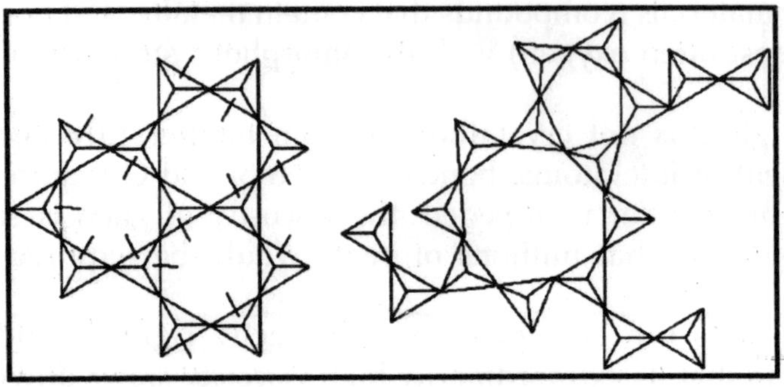

Fig. Two-dimensional View of Tetrahedra as in Silicondioxide Structures: (a) In Crystalline, Ordered Structures and (b) In Disordered Glassy (Amorphous) Structures.

These two phenomena are responsible for the increase in viscosity and the shrinkage that takes place during the cooling period. Beyond the glass transformation temperature changes in configuration of the tetrahedra virtually cease because the glass, to all intents and purposes, has lost its fluidity.

OXIDE GLASSES

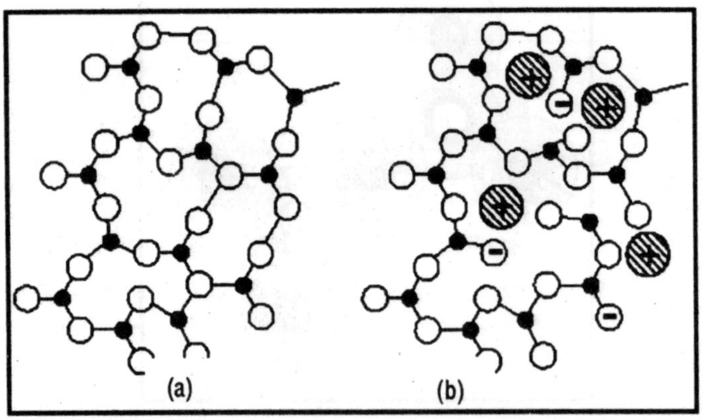

Fig. Schematic Glass Networks; (a) SiO_2 Network (b) SiO_2 network modified through Addition of Na_2O.

The bonding forces responsible for the formation of network structure in crystalline SiO_2 produce similar networks in glasses of other oxides such as B_2O_3 and P_2O_5 which are thus also considered as primary network formers (the present discussion will be limited to silica glasses). If an oxide, such as Na_2O, is added to silica glass, a bond in the network is broken and the relatively mobile sodium ion becomes a part of the structure. Examples of such network modifiers are N_2O, K_2O, Li_2O, CaO, MgO and PbO.

With increase in the amount of modifier, the average number of oxygen-silicon bonds forming bridges between silicon atoms decreases as follows:

O/Si Ratio in Glass	Bridging-O-Si Bonds/Si atom
2.0	4
2.5	3
3.0	2

Two bridging bonds per silicon atom will correspond, for example, to a linear chain structure, and any further increase in modifier would reduce the length of the chain. The principal effect of a modifier is to lower the melting and working temperature by decreasing the viscosity.

An excess of modifier can make the structural units in the melt sufficiently simple and mobile that crystallization occurs in preference to the formation of a glass.

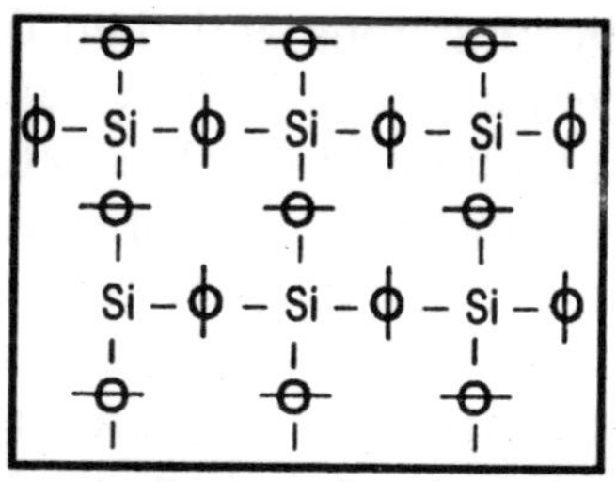

Fig. Schematic of SiO2 Network; The O to Si Tatio is 2 : 1 as Indicated by the Chemical Formula.

The compositions and properties of typical glasses are listed in Table. Soda-lime glass not only has the advantage of being made from inexpensive raw materials, but also the practical convenience of low working temperatures.

Most glass-forming operations (pressing, drawing, etc.) begin at the temperature corresponding to the working point, but can continue while the glass cools.

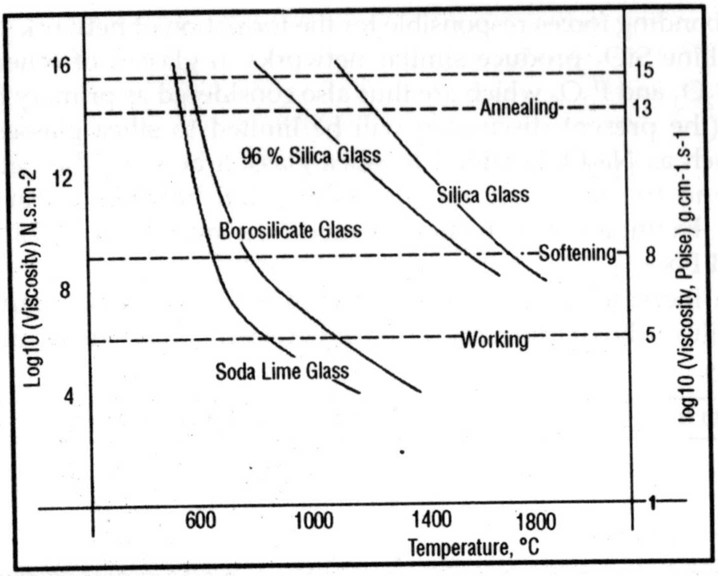

Fig. Important Temperatures in Fabrication of Silicate Glasses (The Temperatures are related to viscosity which in turn controls workability).

ANNEALING OF GLASS

In order to relieve the internal stresses that may develop during glass formation, all commercial glassware is annealed (slowly cooled) immediately after shaping. This slow cooling operation is carried out in industry by placing the shaped objects on a conveyor belt which transports them first through a reheating zone and then through sections having successively lower temperatures in a furnace called the lehr.

The transformation of a liquid to a glass actually takes place over a range of temperatures at which the glass shrinks (thermal contraction). The glass does not undergo a change from disordered liquid structure to crystal structure as iron does at its transformation temperature; in glass, the transformation temperature is the temperature below which viscosity prevents any further configurational changes. Any contraction beyond the transformation temperature range is due only to the lower kinetic energy of the groupings. The transformation temperature of a given "glass composition" depends on its constituents and upon the rate of cooling. Slow cooling results in a lower transformation range because the tetrahedra will have more time to rearrange (to some degree).

This results in tighter packing of tetrahedra as the mass reaches its transformation range. When the glass reaches room temperature, its volume will be smaller when cooled slowly than glass melt which has been cooled rapidly. Hence, slower cooling from the melt results in a denser glass. Glass is thermodynamically unstable at room temperature because configuration changes (which require eons of years to come about!) would result in a lower free energy configuration. However, the changes are so slow that they may be considered negligible.

PROPERTIES OF GLASSES

The strong covalent bonds that exist between the atoms of inorganic glasses plus the lack of crystal structure give glass unusual characteristics. Thus, quartz glass is:

(a) Chemically stable;
(b) A poor conductor of heat and electricity;
(c) Inherently transparent;
(d) Inherently strong.

Chemical Stability

All glasses are immune to oxidation (decay) because their atoms are fully oxidized. Glass composed of 100 per cent silica tetrahedra is extraordinarily inert. It resists the action of most acids (the most notable exception is hydrofluoric acid, utilized to etch glass), but it is attacked by strong alkalis. (The bottles containing liquid alkalis on a chemist's shelf have a clouded surface.)

Conductivity

The ability of a material to conduct electricity is dependent upon the presence of electrons in the conduction band, separated in glasses by as much as 10eV from the valence band. Because the electrons in glass are tightly bonded, glass is a very poor conductor of electricity - in fact, the large energy gap makes glass an excellent insulator.

Glass is also a poor conductor of heat. Thus, hot common glass may crack when cold water is poured onto it because the surface exposed to the cold water will shrink while the dimensions of the interior remain unaffected.

Strength

Glass is harder than many types of steel and is also very elastic, as evidenced by the speed and accuracy with which glass marbles rebound when they meet. Glass has "no" crystal structure, and hence the phenomenon of slip cannot take place. This, together with the strong bonding between atoms, gives glass a very high compressive strength and a theoretical tensile strength of about 107 kN/m^2 (significantly higher than that of steel).

(Glass fibers with an actual tensile strength of 4×10^7 kN/m^2 have been produced.) Since glass has a "liquid structure" it may be considered "saturated with dislocations". [If a piece of metal has a very high dislocation density (eg., due to cold working) the dislocations interfere with each other's movements.

Therefore, with increasing dislocation density a piece of metal becomes harder and stronger, but also more brittle.] Since the molecular structures in glass are unable to move, the presence of minute cracks or imperfections in glass permits stress concentrations to localize and exceed the bond strength between atoms - common glass will crack.

Thus, in actual practice, the strength of glass is, by a factor of 100 to 1000, less than the theoretical strength, and glass is brittle. For example, a freshly made electric light bulb may not crack when initially dropped, but its surface becomes damaged and, after it rebounds and strikes the floor a few more times, it breaks.

The inherent high strength of fibrous glass is utilized in the fiberglass sections employed for boats and automobile bodies. Before being pressed into shape fiberglass is mixed with a synthetic resin which serves to protect the fibers from scratching. Since the fibers thus retain their high strength, fiberglass sections are very strong for their weight.

STRENGTHENED GLASS

The glass scientist has produced objects made of glass which can be struck with a hammer or dropped from a tall building without breaking. Two techniques are employed in strengthening glass - one physical and one chemical. Both techniques are based on the fact that, when there is the slightest imperfection (minute scratch) on the surface, glass remains extraordinarily strong in compression but becomes weak in tension. Thus the strengthening treatment consists of prestressing a glass object by inducing compressive strains in its exterior and thereby enabling it to counteract any tensile stresses which develop under tension.

In physical prestressing, the shaped object is heated to just below its softening point (surface flows are annealed out) and its surface is then chilled by means of a blast of air or an oil bath. Under these conditions the exterior of the glass cools and contracts immediately, but since glass is a poor conductor of heat the interior will not contract - the glass will not crack because the interior remains plastic.

When the interior starts to cool, it cannot contract because the exterior has already set, but, in attempting to contract, the interior continues to draw the exterior together. A built-in compressive stress then develops in the outer layers of the glass. When a tensile stress is applied to this glass, it is counteracted by these compressive stresses. Thus the prestressed glass will not shatter until the surface compression is exceeded. As long as its surface remains intact, prestressed glass continues to be strong and shock-resistant. However, if a deep scratch develops on the surface, the tensile stresses set up in the interior by the surface compression are released and the glass will shatter into thousands of tiny particles. Therefore prestressed articles must be cut to exact size and all holes drilled prior to prestressing. Among the many uses for this type of glass are plate-glass doors, side windows of automobiles and portholes of ships.

Ion exchange is a chemical technique for prestressing glass objects. The first step is to place the shaped object into a molten salt bath containing potassium ions, which replace the sodium ions on the surface of the glass. The potassium ions are larger than the sodium ions and, in accommodating them, the surface of the glass becomes more crowded - thus inducing compressive strains on the exterior.

Glass of this type, characterized by very high strength, is useful for a great variety of applications. Its flexural strength can be as high as 10^6 kN/m^2 compared with 10^5 kN/m^2 for untreated glass.

DEVITRIFICATION OF GLASS

Under certain conditions glass will become "contaminated" with crystalline particles. A glass in this condition is termed devitrified, which is simply another way of stating that it has partially or completely crystallized.

Devitrified glass, unless of the very special variety that is deliberately produced, is undesirable since the crystalline areas form large crystals and are extremely weak and brittle, as well as being translucent only. Crystalline segregations in glass are known as stones.

RECRYSTALLIZED GLASS (GLASS CERAMICS)

Recrystallized glass, also known as polycrystalline glass, is commonly produced by adding nucleating agents to the glass batch. Subsequently the glass can be formed into a desired shape by any of the conventional glass forming processes and then be heat-treated to promote recrystallization.

Recrystallized glass possesses increased impact strength, hardness and thermal shock resistance compared with conventional non-crystalline glasses. One commonplace application of a recrystallized glass is in the manufacture of the so-called refrigerator-to-oven cooking dishes.

COLORED GLASSES

We recognize that glass is normally colorless due to the fact that all electrons are "tightly bonded" and no electronic excitations in the energy range of the visible light spectrum are possible. Glass, however, may be made to absorb selectively in the visible spectrum (thus becoming colored) by any one of three processes:

- *Addition of ions of transition metals*: Such ions provide electronic excitation possibilities to visible light. Typical ions added are:

Cr++	blue
Cr+++	green
Co++	pink
Mn++	orange
Fe++	blue-green

- *Addition of colloidal particles* (s40 to 2000Å): These small particles, while not impeding light transmission (because of their small size) exhibit selective absorption and complementary reflection. Au particles (at 10–4 g/cm3) give colors which vary with particle size:

Au ϕ4-10 nm	pink
Au ϕ10-75 nm	ruby
Au ϕ75-110 nm	green
Au ϕ110-170 nm	brown

- Addition of colored crystals. Some glasses are colored by means of crystals dispersed throughout. An example is the scarlet glass made by the early Egyptians (brought about by the addition of red copper oxide). Another example is red colour from the addition of red Pb_2CrO_6 whereas Cr_2O_3 yields green colour.

GLASS FIBRE

Glass fibre, or fiberglass as it is commonly known, is glass in fibre form. It is made by one of several processes, each of which involves the drawing out of the filaments from glass in the viscous state. Modern technical developments date from World War I when Germany had to find a substitute for asbestos as insulating material.

The process whereby continuous filament glass fibre is made is particularly interesting. The glass batch is melted in the glass tank furnace and the molten glass is molded into ordinary sized marbles. These marbles are fed into an electric furnace which has a platinum-alloy bushing containing many tiny holes.

When the glass marbles reach the holes, the glass flows through them and is drawn down vertically, forming individual filaments. These filaments can then be twisted into a "yarn" and spun on normal textile spindles as required. About one third of an ounce of glass produces about one hundred miles of filament at a rate of about 6,000 feet per minute. Glass wool, on the other hand, is produced by a different process. The most common technique, known as the Crown process, produces a thick matte of short glass fibers which are held together by a polymer binder. In the Crown method, a thick stream of molten glass is poured into a rapidly rotating steel dish that has tiny vents around its periphery. The glass is forced through these vents by centrifugal force, forming relatively short fibers with diameters of about 0.0007 cm.

The matte is then passed through curing ovens where the binder sets and is then cut into sizes suitable for insulation.

SHEET AND PLATE GLASS PRODUCTION

The main types of flat glass produced are drawn sheet and float glass. The processes involved are highly automated.

Flat drawn glass is used for windows and other applications where accuracy of thickness and a high surface finish are not vital. In this process the molten glass is drawn up vertically from the end of a large rectangular tank furnace, the thickness of the drawn sheet being controlled principally by the size of the slot in the steel form held just above the surface of the molten glass. A steel starting dummy is used to commence the drawing action, but this is cracked off once it passes through. Once started, this process is continuous, the cooled and solidified sheet merely being cracked off in suitable lengths once it leaves the rolls.

The float process is the most recent advance in the production of plate glass with a high surface finish. In the float process, flat, fire-finished, stress-free glass is formed by preventing contact between the glass sheet and anything solid. The glass leaves the glass-tank in a continuous molten strip and is floated immediately onto the surface of a bath of molten tin.

The tin bath is surrounded by a non-oxidizing atmosphere and, while in this atmosphere, the glass is heated sufficiently to prevent the development of internal stresses. As the glass sheet leaves the float bath chamber it enters an annealing furnace. Glass produced by the float process is free from distortion and has fire-finished surfaces that are smoother and flatter than those of normal polished plate glass.

Capter 5

Atomic Structure

DISCOVERY OF β, α & γ PARTICLES

The discovery of x-rays by William Conrad Roentgen in November of 1895 excited the imagination of a generation of scientists who rushed to study this phenomenon. Within a few months, Henri Becquerel found that both uranium metal and salts of this element gave off a different form of radiation, which could also pass through solids.

By 1898, Marie Curie found that compoun ds of thorium were also "radioactive." After pain-staking effort she eventually isolated two more radioactive elements–polonium and radium–from ores that contained uranium.

In 1899 Ernest Rutherford found that there were at least two different forms of radioactivity when he studied the absorption of radioactivity by thin sheets of metal foil. One, which he called alpha (α) particles, were absorbed by metal foil that was a few hundredths of a centimeter thick. The other, beta (β) particles, could pass through 100 times as much metal foil before they became absorbed. Shortly thereafter, a third form of radiation, gamma (γ) rays, was discovered that could penetrate as much as several centimeters of lead.

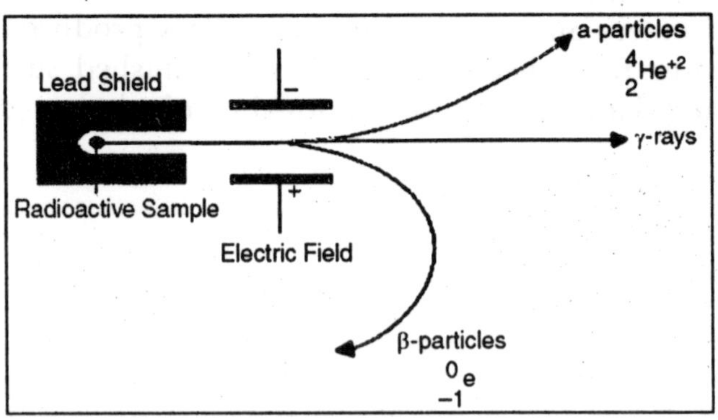

Fig. The Effect of an Electric Field on α-, β, and γ-radiation.

The results of early experiments on these three forms of radiation are shown in the figure below. The direction in which α-particles were deflected by an electric field suggested that they were positively charged. The magnitude of this deflection suggested that they had the same charge-to-mass ratio as an He^{2+} ion.

To test the equivalence between α-particles and He^{2+} ions, Rutherford built an apparatus that allowed α-particles to pass through a very thin glass wall into an evacuated flask that contained a pair of metal electrodes. After a few days, he connected these electrodes to a battery and noted that the gas in the flask did indeed give off the characteristic emission spectrum of helium.

Table. Properties of Radioactive Particles

Name	Symbols	Identity	Relative Charge	Relative Mass	Penetrating Power	Interaction with Charged Plates	Hazards
alpha	α, ^4_2He	helium nucleus	2+	4	low - stopped by a sheet of paper	attracted to negative plate, deflected by positive plate	harmful if ingested
beta	β, $^0_{-1}e$	electron	1–	$\frac{1}{2000}$	moderate - pass through paper and ½mm aluminium, stopped by ½mm lead	attracted to positive plate, deflected by negative plate	skin burns, harmful if ingested particularly iodine-131 in thyroid & strontium-90 in bones
gamma	γ	electromagnetic radiation	0	0	high - only stopped by several centimetres of lead or many centimetres of concrete	unaffected	most dangerous as these are the most penetrating, as a consequence, gamma rays can be used to sterilize materials & destroy bacteria in food

Experiments with electric and magnetic fields demonstrated that β particles were negatively charged. Furthermore, they had the same charge-to-mass ratio as an electron. To date, no detectable difference has been found between β -particles and electrons. The only reason to retain the name " β -particle" is to emphasize the fact that these particles are ejected from the nucleus of an atom when it undergoes radioactive decay.

The fact that γ-rays are not deflected by either electric or magnetic fields suggests that these rays don't carry an electric charge. Since they travel at the speed of light, they are classified as a form of electromagnetic radiation that carries even more energy than x-rays.

GROWTH AND DECAY CURVES

At the turn of the century, when radioactivity was discovered, atoms were assumed to be indestructible. Ernest Rutherford and Frederick Soddy, however, found that radioactive substances became less active with time, as shown in the figure below.

More importantly, they noticed that radioactivity was always accompanied by the formation of atoms of a different element. By 1903, they concluded that radioactivity was accompanied by a change in the structure of the atom. They therefore assumed that radiation was emitted when an element decayed into a different kind of atom.

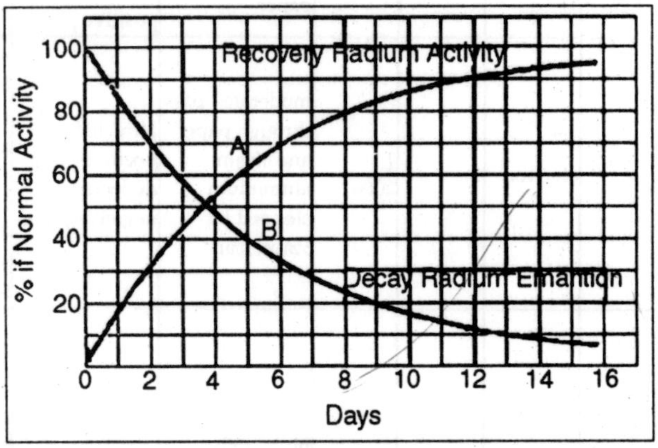

Fig. Growth and Decay Curves Reported by Rutherford and Soddy for "Uranium X" Produced when Uranium Undergoes Radioactive Decay.

By 1910, 40 radioactive elements had been isolated that were associated with the process by which uranium metal decayed to lead. This created a problem, however, because there was space for only 11 elements between lead and uranium. In 1913, Kasimir Fajans and Fredick Soddy proposed an explanation for these results based on the following rules.

- α-particles are emitted when an element is formed that belongs two spaces to the left in the periodic table. Uranium (Z = 92), for example, emits an α-particle when it decays to form thorium (Z = 90).

- β -particles are emitted when an element is formed that belongs one space to the right in the periodic table. Actinium (Z = 89) emits a β -particle when it decays to form thorium (Z = 90).
- Radioactive elements that fall in the same place in the periodic table are different forms of the same element. The radioactive thorium produced by the α-particle decay of uranium is a different form of the element than the radioactive thorium obtained by the β -particle decay of actinium.

Soddy proposed the name isotope to describe different radioactive atoms that occupy the same position in the periodic table. J. J. Thomson and Francis Aston then used a mass spectrometer to show that isotopes are atoms of the same element that have different atomic masses.

THE STRUCTURE OF THE ATOM

The discovery of the electron in 1897 by J. J. Thomson suggested that there was an internal structure to the "indivisible" building blocks of matter known as atoms. This raised an obvious question: How many electrons does an atom contain? By studying the scattering of light, x-rays, and α-particles, Thomson concluded that the number of electrons in an atom was between 0.2 and 2 times the weight of the atom.

In 1911, Rutherford concluded that the scattering of α-particles by extremely thin pieces of metal foil could be explained by assuming that all of the positive charge and most of the mass of the atom were concentrated in an infinitesimally small fraction of the total volume of the atom, for which he proposed the name nucleus. Rutherford's data also suggested that the nucleus of a gold atom carries a positive charge that is about 80 times the charge on an electron.

The discovery of the neutron in 1932 explained the discrepancy between the charge on the nucleus and the mass of an atom. A neutral gold atom that has a mass of 197 amu consists of a nucleus that contains 79 protons and 118 neutrons surrounded by 79 electrons. By convention, this information is specified by the following symbol, which describes the only naturally occurring isotope of gold.

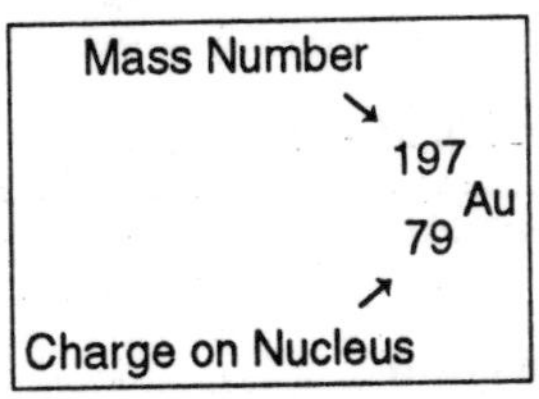

This convention can also be applied to subatomic particles. The only difference is the use of lowercase letters to identify the particle.

$$\underset{\text{electron}}{{}^{0}_{1}e} \quad \underset{\text{Proton}}{{}^{1}_{1}P} \quad \underset{\text{Neutron}}{{}^{1}_{0}n}$$

Because anyone with access to a periodic table can find the atomic number of an element, a shorthand notation is often used that reports only the mass number of the atom and the symbol of the element. The shorthand notation for the naturally occurring isotope of gold is ^{197}Au.

RADIOACTIVE DECAY

α, β & γ Radiation

Early studies of radioactivity indicated that three different kinds of radiation were emitted, symbolized by the first three letters of the Greek alphabet α, β and α. With time, it became apparent that this classification scheme was much too simple. The emission of a negatively charged β particle, for example, is only one example of a family of radioactive transformations known as *β-decay. A fourth category, known as spontaneous fission, also had to be added to describe the process by which certain radioactive nuclides decompose into fragments of different weight.*

ALPHA DECAY

Alpha decay is usually restricted to the heavier elements in the periodic table. (Only a handful of nuclides with atomic numbers less than 83 emit an α-particle.) The product of α-decay is easy to predict if we assume that both mass and charge are conserved in nuclear reactions. Alpha decay of the ^{238}U "parent" nuclide, for example, produces ^{234}Th as the "daughter" nuclide.

$$^{238}_{92}U \rightarrow {}^{234}_{90}Th + {}^{4}_{2}He(\alpha - \text{decay})$$

The sum of the mass numbers of the products (234 + 4) is equal to the mass number of the parent nuclide (238), and the sum of the charges on the products (90 + 2) is equal to the charge on the parent nuclide.

BETA DECAY

There are three different modes of beta decay:

- Electron (β^-) emission
- Electron capture
- Positron (β^+) emission

Electron (β^-) emission is literally the process in which an electron is ejected or emitted from the nucleus. When this happens, the charge on the nucleus increases by one. Electron (β^-) emitters are found throughout the periodic table, from the lightest elements (^3H) to the heaviest (^{255}Es). The product of β^--emission can be predicted by assuming that both mass number and charge are conserved in nuclear reactions. If ^{40}K is a β^--emitter, for example, the product of this reaction must be ^{40}Ca.

$$^{40}_{19}K \rightarrow {}^{40}_{20}Ca + {}^{0}_{-1}e[\text{electron}(\beta^-)\text{emission}]$$

Once again the sum of the mass numbers of the products is equal to the mass number of the parent nuclide and the sum of the charge on the products is equal to the charge on the parent nuclide.

Nuclei can also decay by capturing one of the electrons that surround the nucleus. Electron capture leads to a decrease of one in the charge on the nucleus. The energy given off in this reaction is carried by an x-ray photon, which is represented by the symbol hv, where h is Planck's constant and v is the frequency of the x-ray.

The product of this reaction can be predicted, once again, by assuming that mass and charge are conserved.

$$^{40}_{19}K + {}^{0}_{-1}a \rightarrow {}^{40}_{81}Ar + hv[\text{electron Capture}]$$

The electron captured by the nucleus in this reaction is usually a 1s electron because electrons in this orbital are the closest to the nucleus.

A third form of beta decay is called positron (β^+) emission. The positron is the antimatter equivalent of an electron. It has the same mass as an electron, but the opposite charge. Positron (β^+) decay produces a daughter nuclide with one less positive charge on the nucleus than the parent.

$$^{40}_{19}K \rightarrow {}^{40}_{81}Ar \rightarrow {}^{0}_{1}e[\text{Positron}(\beta)\text{emossion}]$$

Positrons have a very short life-time. They rapidly lose their kinetic energy as they pass through matter. As soon as they come to rest, they combine with an electron to form two γ-ray photons in a matter-antimatter annihilation reaction.

$$^{0}_{1}e \rightarrow {}^{0}_{-1}e \rightarrow e \rightarrow 2\gamma$$

Thus, although it is theoretically possible to observe a fourth mode of beta decay corresponding to the capture of a positron, this reaction does not occur in nature. Note that in all three forms of β -decay for the ^{40}K nuclide the mass number of the parent and daughter nuclides are the same for electron emission, electron capture, and position emission. All three forms of β -decay therefore interconvert isobars.

GAMMA EMISSION

The daughter nuclides produced by β-decay or β -decay are often obtained in an excited state. The excess energy associated with this excited state is released when the nucleus emits a photon in the γ-ray portion of the electromagnetic spectrum.

Most of the time, the γ-ray is emitted within 10^{-12} seconds after the α-particle or β -particle. In some cases, gamma decay is delayed, and a short-lived, or metastable, nuclide is formed, which is identified by a small letter m written after the mass number. ^{60m}Co, for example, is produced by the electron emission of ^{60}Fe.

$$^{60}_{26}\mathrm{Fe} \rightarrow {}^{60m}_{27}\mathrm{Co} + {}^{0}_{-1}\mathrm{e}$$

The metastable ^{60m}Co nuclide has a half-life of 10.5 minutes. Since electromagnetic radiation carries neither charge nor mass, the product of γ-ray emission by ^{60m}Co is ^{60}Co.

$$^{60}_{26}\mathrm{Co} \rightarrow {}^{60}_{27}\mathrm{Co} + \gamma(\gamma - \mathrm{ray\,emossion}]$$

SPONTANEOUS FISSION

Nuclides with atomic numbers of 90 or more undergo a form of radioactive decay known as spontaneous fission in which the parent nucleus splits into a pair of smaller nuclei. The reaction is usually accompanied by the ejection of one or more neutrons.

$$^{252}_{98}\mathrm{Cf} \rightarrow {}^{140}_{54}\mathrm{Xe} + {}^{108}_{44}\mathrm{Fu} + 4\,{}^{1}_{0}\mathrm{n}$$

For all but the very heaviest isotopes, spontaneous fission is a very slow reaction. Spontaneous fission of ^{238}U, for example, is almost two million times slower than the rate at which this nuclide undergoes α-decay.

NEUTRON-RICH VERSUS NEUTRON-POOR NUCLIDES

In 1934 Enrico Fermi proposed a theory that explained the three forms of beta decay. He argued that a neutron could decay to form a proton by emitting an electron. A proton, on the other hand, could be transformed into a neutron by two pathways. It can capture an electron or it can emit a positron. Electron emission therefore leads to an increase in the atomic number of the nucleus.

$$^{14}_{6}C \rightarrow ^{14}_{7}N + ^{0}_{-1}e$$

Both electron capture and positron emission, on the other hand, result in a decrease in the atomic number of the nucleus.

$$^{7}_{4}Be + ^{0}_{-1}e \rightarrow ^{7}_{3}Li + hv$$

$$^{11}_{6}C \rightarrow ^{11}_{5}B + ^{0}_{+1}e$$

A plot of the number of neutrons versus the number of protons for all of the stable naturally occurring isotopes is shown in the figure below. Several conclusions can be drawn from this plot.

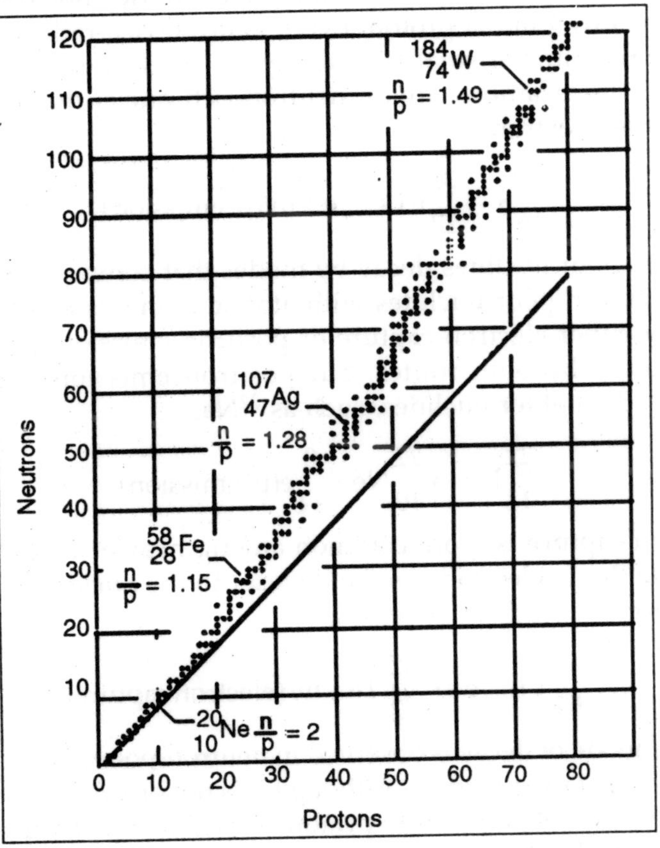

Fig. A Graph of the Number of Neutrons Versus the Number of Protons for all Stable Naturally Occurring nuclei.

- The stable nuclides lie in a very narrow band of neutron-to-proton ratios.

- The ratio of neutrons to protons in stable nuclides gradually increases as the number of protons in the nucleus increases.
- Light nuclides, such as ^{12}C, contain about the same number of neutrons and protons. Heavy nuclides, such as ^{238}U, contain up to 1.6 times as many neutrons as protons.
- There are no stable nuclides with atomic numbers larger than 83.
- This narrow band of stable nuclei is surrounded by a sea of instability.
- Nuclei that lie above this line have too many neutrons and are therefore neutron-rich.
- Nuclei that lie below this line don't have enough neutrons and are therefore neutron-poor.

The most likely mode of decay for a neutron-rich nucleus is one that converts a neutron into a proton. Every neutron-rich radioactive isotope with an atomic number smaller 83 decays by electron (β^-) emission. ^{14}C, ^{32}P, and ^{35}S, for example, are all neutron-rich nuclei that decay by the emission of an electron.

$$^{35}_{16}S \rightarrow ^{35}_{17}Cl + ^{0}_{+1}e + h\nu(\beta^- \text{emission})$$

Neutron-poor nuclides decay by modes that convert a proton into a neutron. Neutron-poor nuclides with atomic numbers less than 83 tend to decay by either electron capture or positron emission. Many of these nuclides decay by both routes, but positron emission is more often observed in the lighter nuclides, such as ^{22}Na.

$$^{22}_{11}Na \rightarrow ^{22}_{10}Ne + ^{0}_{+1}e(\beta^+ \text{emission})$$

Electron capture is more common among heavier nuclides, such as ^{125}I, because the 1s electrons are held closer to the nucleus of an atom as the charge on the nucleus increases.

$$^{125}_{53}I + ^{0}_{-1}e \rightarrow ^{125}_{52}Te + h\nu(\text{electron capture})$$

A third mode of decay is observed in neutron-poor nuclides that have atomic numbers larger than 83. Although it is not obvious at first, α-decay increases the ratio of neutrons to protons. Consider what happens during the α-decay of ^{238}U, for example.

$$^{238}_{92}U \rightarrow ^{234}_{90}Th + ^{4}_{2}He(\alpha \text{ decay})$$

The parent nuclide (^{238}U) in this reaction has 92 protons and 146 neutrons, which means that the neutron-to-proton ratio is 1.587. The daughter nuclide (^{234}Th) has 90 protons and 144 neutrons, so its neutron-to-proton ratio is 1.600. The daughter nuclide is therefore slightly less likely to be neutron-poor, as shown in the figure below.

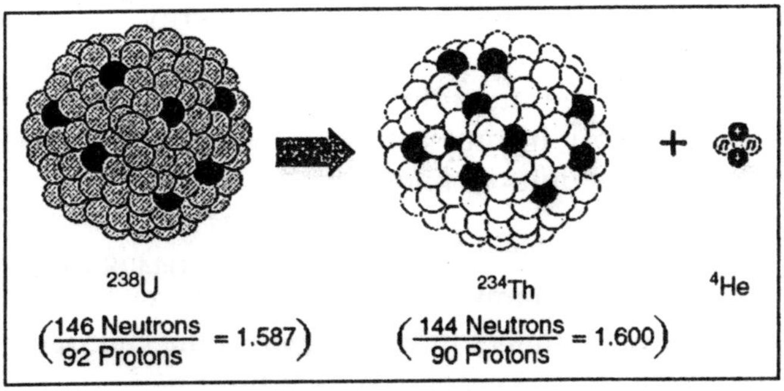

BINDING ENERGY CALCULATIONS

We should be able to predict the mass of an atom from the masses of the subatomic particles it contains. A helium atom, for example, contains two protons, two neutrons, and two electrons.

Example: The mass of a helium atom should be 4.0329802 amu.

2(1.0072765) amu	=	2.0145530 amu
2(1.0086650) amu	=	2.0173300 amu
2(0.0005486) amu	=	0.0010972 amu
Total mass	=	4.0329802 amu

When the mass of a helium atom is measured, we find that the experimental value is smaller than the predicted mass by 0.0303769 amu.

Predicted mass	=	4.0329802 amu
Observed mass	=	4.0026033 amu
Mass defect	=	0.0303769 amu

The difference between the mass of an atom and the sum of the masses of its protons, neutrons, and electrons is called the mass defect. The mass defect of an atom reflects the stability of the nucleus. It is equal to the energy released when the nucleus is formed from its protons and neutrons. The mass defect is therefore also known as the binding energy of the nucleus.

The binding energy serves the same function for nuclear reactions as ΔH for a chemical reaction. It measures the difference between the stability of the products of the reaction and the starting materials. The larger the binding energy, the more stable the nucleus. The binding energy can also be viewed as the amount of energy it would take to rip the nucleus apart to form isolated neutrons and protons. It is therefore literally the energy that binds together the neutrons and protons in the nucleus.

The binding energy of a nuclide can be calculated from its mass defect with Einstein's equation that relates mass and energy.

$E = mc^2$

Example: We found the mass defect of He to be 0.0303769 amu. To obtain the binding energy in units of joules, we must convert the mass defect from atomic mass units to kilograms.

$$0.030769\text{amu} \times \frac{1.6605655 \times 10^{-24}\text{g}}{1\text{amu}} \times \frac{1\text{kg}}{1000\text{g}} = 5.04428 \times 10^{-29}\text{kg}$$

Multiplying the mass defect in kilograms by the square of the speed of light in units of meters per second gives a binding energy for a single helium atom of 4.53358×10^{-12} joules.

$$\text{E} \times (5.04428 \times 10^{-29}\text{kg})(2.9979246 \times 10^{8}\text{m/s})^2 = 4.53358 \times 10^{-12}\text{f}$$

Multiplying the result of this calculation by the number of atoms in a mole gives a binding energy for helium of 2.730×10^{12} joules per mole, or 2.730 billion kilojoules per mole.

$$\frac{4.538 \times 10^{-12}\text{J}}{1\text{atom}} \times \frac{6.022 \times 10^{23}\text{atoms}}{1\text{mol}} = 2.730 \times 10^{-12}\text{J/mol}$$

This calculation helps us understand the fascination of nuclear reactions. The energy released when natural gas is burned is about 800 kJ/mol. The synthesis of a mole of helium releases 3.4 million times as much energy.

Since most nuclear reactions are carried out on very small samples of material, the mole is not a reasonable basis of measurement. Binding energies are usually expressed in units of electron volts (eV) or million electron volts (MeV) per atom.

Example: The binding energy of helium is 28.3 x 10^6 eV/atom or 28.3 MeV/atom.

$$\frac{4.53358 \times 10^{-12}\text{J}}{1\text{atom}} \times \frac{1\text{ev}}{1.6021892 \times 10^{-19}\text{J}} = 28.30 \times 10^{6}\text{eviatom}$$

Calculations of the binding energy can be simplified by using the following conversion factor between the mass defect in atomic mass units and the binding energy in million electron volts.

1 amu = 931.5016 MeV

Binding energies gradually increase with atomic number, although they tend to level off near the end of the periodic table. A more useful quantity is obtained by dividing the binding energy for a nuclide by the total number of protons and neutrons it contains. This quantity is known as the binding energy per nucleon.

The binding energy per nucleon ranges from about 7.5 to 8.8 MeV for most nuclei, as shown in the figure below. It reaches a maximum, however, at an atomic mass of about 60 amu. The largest binding energy per nucleon is observed for ^{56}Fe, which is the most stable nuclide in the periodic table.

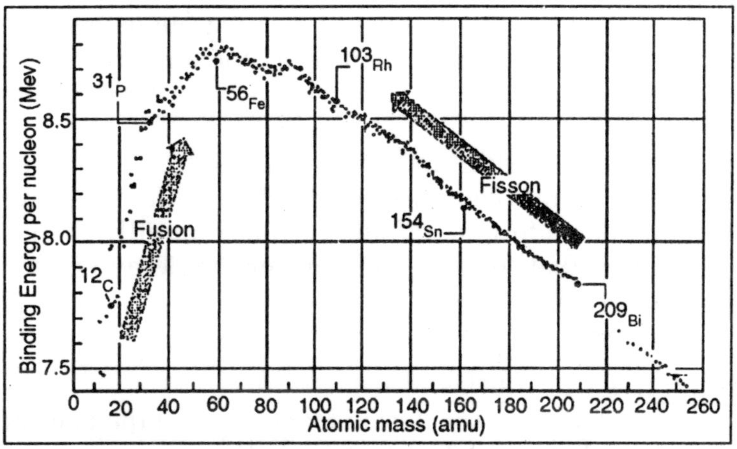

The graph of binding energy per nucleon versus atomic mass explains why energy is released when relatively small nuclei combine to form larger nuclei in fusion reactions.

$$^{12}_{6}C + ^{24}_{12}C \rightarrow ^{24}_{12}Mg(\text{fusion})$$

It also explains why energy is released when relatively heavy nuclei split apart in fission (literally, "to split or cleave") reactions.

$$^{235}_{92}U \rightarrow ^{139}_{56}Ba + ^{94}_{36}Kr + 2\,^{1}_{0}n(\text{fusion})$$

There are a number of small irregularities in the binding energy curve at the low end of the mass spectrum, as shown in the figure below. The ^4He nucleus, for example, is much more stable than its nearest neighbors. The unusual stability of the ^4He nucleus explains why α-particle decay is usually much faster than the spontaneous fission of a nuclide into two large fragments.

THE KINETICS OF RADIOACTIVE DECAY

Radioactive nuclei decay by first-order kinetics. The rate of radioactive decay is therefore the product of a rate constant (k) times the number of atoms of the isotope in the sample (N).

$$\text{Rate} = kN$$

The rate of radioactive decay doesn't depend on the chemical state of the isotope. The rate of decay of ^{238}U, for example, is exactly the same in uranium metal and uranium hexafluoride, or any other compound of this element.

The rate at which a radioactive isotope decays is called the activity of the isotope. The most common unit of activity is the curie (Ci), which was originally defined as the number of disintegrations per second in 1 gram of ^{226}Ra. The curie is now defined as the amount of radioactive isotope necessary to achieve an activity of 3.700 x 10^{10} disintegrations per second.

The relative rates at which radioactive nuclei decay can be expressed in terms of either the rate constants for the decay or the half-lives of the nuclei. We can conclude that ^{14}C decays more rapidly than ^{238}U, for example, by noting that the rate constant for the decay of ^{14}C is much larger than that for ^{238}U.

^{14}C: $k = 1.210 \times 10^{-4}\ y^{-1}$

^{238}U: $k = 1.54 \times 10^{-10}\ y^{-1}$

We can reach the same conclusion by noting that the half-life for the decay of ^{14}C is much shorter than that for ^{235}U.

^{14}C: $t_{1/2} = 5730$ y

^{238}U: $t_{1/2} = 4.51 \times 10^{9}$ y

The half-life for the decay of a radioactive nuclide is the length of time it takes for exactly half of the nuclei in the sample to decay. In our discussion of the kinetics of chemical reactions, we concluded that the half-life of a first-order process is inversely proportional to the rate constant for this process.

$$t_{1/2} = \frac{\text{In}2}{k} = \frac{0.693}{k}$$

The half-life of a nuclide can be used to estimate the amount of a radioactive isotope left after a given number of half-lives. For more complex calculations, it is easier to convert the half-life of the nuclide into a rate constant and then use the integrated form of the first-order rate law described in the kinetic section.

DATING BY RADIOACTIVE DECAY

The earth is constantly bombarded by cosmic rays emitted by the sun. The total energy received in the form of cosmic rays is small –no more than the energy received by the planet from starlight. But the energy of a single cosmic ray is very large, on the order of several billion electron volts (0?200 million kJ/mol). These highly energetic rays react with atoms in the atmosphere to produce neutrons that then react with nitrogen atoms in the atmosphere to produce ^{14}C.

$$^{14}_{7}N + ^{1}_{0}n \rightarrow ^{14}_{6}C + ^{1}_{1}H$$

The ^{14}C formed in this reaction is a neutron-rich nuclide that decays by electron emission with a half-life of 5730 years.

$$^{14}_{6}C \rightarrow ^{14}_{7}N + ^{0}_{-1}e$$

Just after World War II, Willard F. Libby proposed a way to use these reactions to estimate the age of carbon-containing substances. The ^{14}C dating technique for which Libby received the Nobel prize was based on the following assumptions.

- ^{14}C is produced in the atmosphere at a more or less constant rate.
- Carbon atoms circulate between the atmosphere, the oceans, and living organisms at a rate very much faster than they decay. As a result, there is a constant concentration of ^{14}C in all living things.
- After death, organisms no longer pick up ^{14}C.

Thus, by comparing the activity of a sample with the activity of living tissue we can estimate how long it has been since the organism died.

The natural abundance of ^{14}C is about 1 part in 10^{12} and the average activity of living tissue is 15.3 disintegrations per minute per gram of carbon. Samples used for ^{14}C dating can include charcoal, wood, cloth, paper, sea shells, limestone, flesh, hair, soil, peat, and bone. Since most iron samples also contain carbon, it is possible to estimate the time since iron was last fired by analyzing for ^{14}C.

We now know that one of Libby's assumptions is questionable: The amount of ^{14}C in the atmosphere hasn't been constant with time. Because of changes in solar activity and the earth's magnetic field, it has varied by as much as 5%. More recently, contamination from the burning of fossil fuels and the testing of nuclear weapons has caused significant changes in the amount of radioactive carbon in the atmosphere. Radiocarbon dates are therefore reported in years before the present era (B.P.). By convention, the present era is assumed to begin in 1950, when ^{14}C dating was introduced.

Studies of bristlecone pines allow us to correct for changes in the abundance of ^{14}C with time. These remarkable trees, which grow in the White Mountains of California, can live for up to five thousand years. By studying the ^{14}C activity of samples taken from the annual growth rings in these trees, researchers have developed a calibration curve for ^{14}C dates from the present back to 5145 B.C. After roughly 45,000 years (eight half-lives), a sample retains only 0.4% of the ^{14}C activity of living tissue. At that point it becomes too old to date by radiocarbon techniques. Other radioactive isotopes can be used to date rocks, soils, or archaeological objects that are much older.

Potassium-argon dating, for example, has been used to date samples up to 4.3 billion years old. Naturally occurring potassium contains 0.0118% by weight of the radioactive ^{40}K isotope.

This isotope decays to ^{40}Ar with a half-life of 1.3 billion years. The ^{40}Ar produced after a rock crystallizes is trapped in the crystal lattice. It can be released, however, when the rock is melted at temperatures up to 2000C. By measuring the amount of ^{40}Ar released when the rock is melted and comparing it with the amount of potassium in the sample, the time since the rock crystallized can be determined.

Chapter 6

Composite Structure of an Atom

BOHR'S MODEL

Structure of Atom

The Danish physicist Niels Henrik David Bohr has created the theory of a structure of atom, which was deprived of the basic deficiencies of Rutherford's model of atom. Thus he had to change some habitual representations of classical physics.

Bohr has formulated three postulates (sometimes the first and the second postulates are combined, then Bohr's theory of atom is reduced to two postulates):

- The electrons in atom rotate around the kern on particular stationary orbits only. The motions on these orbits correspond to stationary states of atom, which do not vary in time without exterior actions. Rotating on stationary orbits, the electrons do not radiate energy.
- Allowed, discrete values of energy of an electron correspond to stationary states of atom.
  This postulate was called the quantization rule of orbits. It affirms, that when the electron is going on a stationary orbit, it should have discrete values of a moment of motion:

$$m_0 ur = n\frac{h}{2\pi}, n = 1, 2, 3, \ldots$$

M_0 -electronic mass; v-velocity of atom:
r-radius of an orbit
h-planck's constant:
$h = 6{,}62 \times 10^{-34}$

- The radiation or uptake of energy by atom occurs only at transition of an electron from one stationary orbit to another. Thus quantum of light energy (photon) is radiated or is immersed. The energy of a quantum is equal to a difference of energies of an electron on the relevant stationary orbits: frequency of a light, energy of an electron on m and n orbits.

The laws of classical physics say, that the transition of bodies from one state to another occurs continuously at continuous radiation or uptake of energy. The classical laws have appeared inapplicable to the phenomena, which are described by Bohr.

Parameters of Atom

The radiuses of stationary orbits and energies of electrons are determined by dependence of energy of atom on radius of an electronic orbit. The elementary system is the hydrogen-like atom. The hydrogen-like atom is the one, in which one electron rotates on a circular orbit around of a kern, which has a charge *Ze* (Z - serial number of a device in periodic system).

The energy of an electron in atom consists of a kinetic energy of a motion on an orbit and potential energy in an electric field of a kern (electronic mass, travelling speed of an electron on an orbit, radius of an orbit, permittivity of vacuum of empty space):

$$E = \frac{m_0 v^2}{2} + \frac{-Ze^2}{4\pi\varepsilon_0 r} = \frac{m_0 v^2}{2} - \frac{Ze^2}{4\pi\varepsilon_0 r}$$

M_0 -electronic mass: u-velocity of atom:

ε_0-permittivity of vacuum of emepty space:

r-radius of an orbit

The kinetic energy is determined with the requirement, that a centripetal acceleration is created by force of a Coulomb attraction of charges. The kinetic energy of an electron is inversely proportional to radius of an orbit. A total energy:

$$E = \frac{1}{2}\frac{-Ze^2}{4\pi\varepsilon_0 r} - \frac{-Ze^2}{4\pi\varepsilon_0 r} = -\frac{1}{2}\frac{Ze^2}{4\pi\varepsilon_0 r}$$

The total energy of an electron varies with the change of a radius of an orbit the same as potential energy does. Hence, the more than radius of an orbit of an electron, which rotates around the kern, the larger is the total energy of atom.

Value $r_1 = 0.528 \times 10^{-10}$ m is called the first Bohr's radius. For the subsequent values: $r_n = n^2 r_1$.

n - main quantum number. It specifies an energy level, on which the atom is. The lowest state of atom is inconvertible. All atoms are in the lowest state at standard conditions.

The atom can be exited, if the energy transferred to it. Thus it will be transferred to one of higher energy states. The distance from the lowest energy level $n = 1$ up to the proximate level with higher energy n = 2 makes 10,1 eV = 16.2×10^{-19} J for the atom of hydrogen. This energy is the least bundle of energy, which the atom of hydrogen can immerse, when it is in a normal state.

Deficiencies of the Theory

Bohr's theory explained the nature of a spectral series of hydrogen. It has allowed to spot theoretically the sizes of atom and energy levels of electrons for atom of hydrogen.

Deficiency of Bohr's theory is the inconsistency of its build-up. Bohr's postulates have quantum character, but the stationary electronic orbits and allowed levels of energy of atom are determined by methods of a classical mechanics and electrodynamics. Therefore Bohr's theory has appeared usable to calculations only of monovalent atoms.

Bohr's theory does not give an opportunity to calculate intensity of spectral lines, which is interlinked to probabilities of electronic transitions. The introduction of quantum rules for definition of orbits and levels of electrons was not logically justified by Bohr.

The further development of the Bohr's theory was made by German physicist Arnold Johannes Wilhelm Sommerfeld has changed circular orbits on elliptic. It has enabled to apply Bohr's theory to multi electronic atoms. However the creation of a fundamental theory of atom is bound to the creation of a quantum mechanics.

RUTHERFORD'S ATOMIC MODEL

The perception of an atom as the least indivisible particle of substance was the foundation of the cinematic gas theory. It have explained and predicted a series of the experimental phenomena. However, at the end of XIX century new experimental data proved the composite structure of an atom.

The experimentally found laws of radiation and uptake of energy by atoms (spectral series, radiation of an ideal black body, photoeffect) have not received an explanation within the framework of a classical electrodynamics. Only new ideas, based on the quantum theory, helped explaining these phenomenon.

Structure of Atom

British physicist Ernest Rutherford has created a planetary model of atom exploring a dispersion of a-particles transiting through a thin metal foil. According to this model an atom looks like a tiny planetary system, in which the forces of an electrical attraction operate.

The centre of an atom is a positively charged kern. Almost all mass of an atom is focused there. Negatively charged eletrons rotate around the kern.

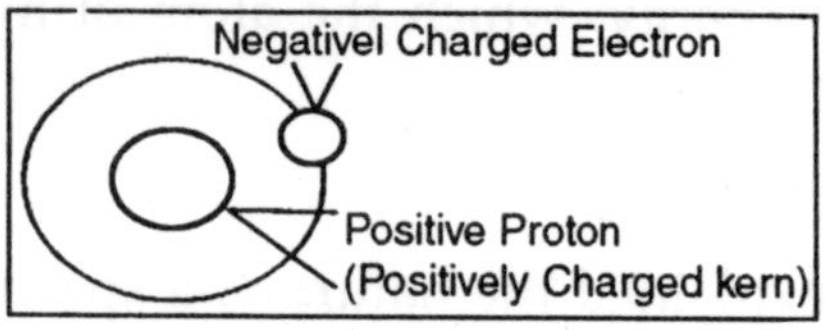

Fig. Structure of Atom.

An atom of hydrogen has the simplest structure. Positively charged kern of an atom of hydrogen was called a proton. The only electron rotates around it. It was determined later, that every atom's kern consists of protons and neutrons. The neutron's mass is almost equal to protons, but has not electric charge. The mass of an electron is about 1/1840 of a proton's mass and does not influence an atomic weight of substance. The number of positive and negative charges of atom determines the position of a substance in a periodic system.

The positive charge of an atom's kern is equal to a serial number of a substance in the table. The number of neutrons in an atom's kern is equal to a residual between a rounded atomic weight and a number of positive protons in a kern (serial number of substance). The electrical neutrality rule of atom says that the number of electrons rotating around the kern is equal to a number of positive protons contained in an atom.

Parameters of Atom

According to the Rutherford's experimental results the radius of a trajectory of an electron in a hydrogen atom is 0.53×10^{-10} m, the diameter of an atom of hydrogen is equal to 2.6×10^{-15} m. Volume of electrons and kern is approximately 1/10 000 of an atom's volume. The substance of a kern has huge density of 200 million ton in 1 cm^3. It is all because the most of an atom's mass is concentrated in its kern. The mass of a positive proton is approximately equal to the mass of a neutron and is 1.67×10^{-27}. The mass of an electron is $9, 11 \times 10^{-31}$ kg.

The classic radius of an electron, r_0, is equal to 2.8×10^{-17} m. It is the distance, on which the electrostatic energy of an electron intercourse with a point charge of the same quantity is equal to natural energy of an electron:

$$\frac{e^2}{r_0} = m_0 c^2$$

e - elementary charge; *c* - speed of light in vacuum:
m_0 - electronic mass

Deficiencies of the Theory

The Rutherford model of atom was well correlated with the results of a dispersion of an a-particle by atoms of substance, but it has not explained neither the process of radiation of atoms, nor legitimacies in spectrums of radiation. According to the laws of an electrodynamics rotating around the kern the electron should radiate electromagnetic waves. As a result of radiation the natural energy of an electron should be diminished, thus the trajectory it will be figured by a spiral, and during the order 10^{-8} about an electron should fall on a kern. Such deduction obtained on the basis of representations of classical physics about radiation, contradicted known stability of atoms and character of nuclear spectrums of radiation.

MULTIELECTRONIC ATOMS

Structure of Atom

Just as in the atom of hydrogen, in more composite atoms electrons can rotate around of a kern on allowed orbits only. The orbits in composite atoms cluster in system of shells. Each shell contains a particular number of orbits, an electron can be on any og them. A *K*-shell is the closest to a kern and contains two orbits. The second is a *L*-shell has eight orbits. The third is a *M*-shell has eight orbits, the fourth *N*-shell has 18 orbits etc.

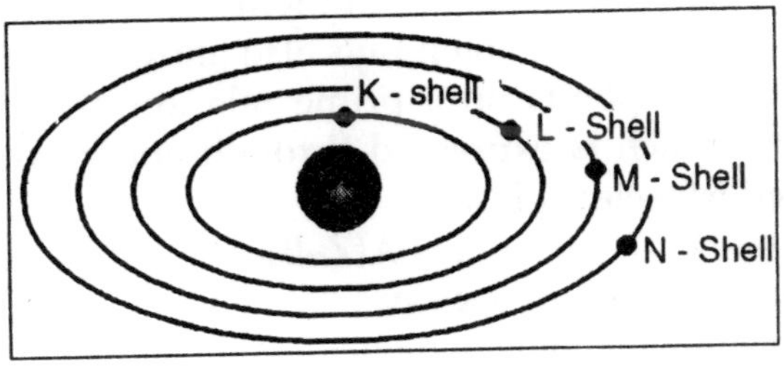

Fig. Structure of Atom.

Parameters of Atom

When the electron travels from an orbit of greater radius to an orbit of smaller radius, the energy is oozed. But the transition from outer shells on interior is possible only when there are free orbits on the leter. The outer-shell electrons are less bound to a kern than interior ones: they are on greater distance from a kern, and the electrons, which are on inner shells, screen the nuclear charge.

Therefore just a small amount of energy is sufficient for excitation of outer-shell electrons. The quantums of energy radiated at transition of such electrons in a nonexcited state have a quantity of about several electron-volt. Greater energy is necessary to detach inner-shell electrons from the atom. The greater the nuclear charge the greater energy is needed.

Therefore, when there is a transition of electrons from outer shells to a *K*-shell in heavy atoms, the emission of quantums with energy of hundreds and even thousand electron-volt is possible. Such quantums of high energy are called Roentgen rays.

X-ray Spectrums of Atoms

The X-rays results from losses by atom of an inner-shell electron: The transition of electrons is accompanied by an emission of quantums of high energy with a small wave length, which are termed as Roentgen rays.

The X-rays emerges when atom losses inner-shell electron: The transition of electrons is accompanied by an emission of quantums of high energy with a small wave length called Roentgen rays.

The atom can lose inner-shell electrons as a result of an electron bombardment with electrons speeded up to energy of several thousand electron-volt. The wave length of the X-rays is comparable to interatomic distance in solid bodies. The examination of the X-ray diffraction on crystal lattices of solid bodies enables, knowing a wave length of Roentgen rays, to determine interatomic distances in them.

The examination of Roentgen rays also allows t3o determine nuclear charge, because it depends on a prime relation of an electron-binding energy in atom, which is expressed through energy of the X-rays. This relation is called Moseley's law:

$$v = A(Z-B)$$

V-frequency of a line in a x-ray
characteristic spectrum of a devic:
A and B-constants: Z-serial number of a device
in periodic system

PROPERTIES OF LIGHT

In one phenomena, which are bound to distribution of light, its undular properties (interference, diffraction) are shown. In other phenomena its quantum properties (photoeffect, radiation and uptake of a light, process of interaction of a light with substance) are shown.

The light has simultaneously undular and quantum properties, which are differently shown in the different phenomena. However, when the wave length of a light is diminished, its undular properties are attenuated and the quantum properties are shown.

Compton's effect

The experience of Arthur Holly Compton on a dispersion of Roentgen rays by substances, which consist of atoms of light devices, have shown, that the dispelled Roentgen rays have a major wave length λ', than impinging λ. Thus the residual $\Delta\lambda = \lambda' - \lambda$ depends on properties of scattering substance and wave length of an impinging light:

$$\Delta\lambda = \lambda' - \lambda = 2\lambda_K \sin^2 \frac{\theta}{2}$$

The quantity λ_K is a stationary value for all substances also is termed as Compton's wave length, $\lambda_K = 2.43 \times 10^{12}$ m. This phenomenon, which is termed as Compton's recoil, is impossible to explain by a wave theory of light. In quantum theory the magnification of wave length of radiation at a dispersion is explained by interaction of photons of impinging radiation with electrons of atoms and molecules.

The part of energy of an impinging photon, which is transmitted to electrons of scattering substance, depends on a scattering angle of photon. At $\theta = \theta$, that is provided that after a dispersion a photon will fly in the party, which is opposite to a tentative direction, the losses by a photon of energy will be maximal, that will give in maximal magnification of a wave length of radiation.

Light Pressure

Some properties, for example light pressure, can be surveyed simultaneously in the electromagnetic theory and in the quantum theory. According to the electromagnetic theory of light, the pressure *p*, which renders a luminous flux with power *W*, impinging normally on a simple surface, is equal:

$$P = \frac{W}{c}(1+R)$$

For an ideal black body (R = 0): $P = \frac{W}{c}$

For an absolute reflecting body (R = 1): $P = \frac{2W}{c}$

If *N* of quantums of a luminous flux with frequency n, which bears(carries) energy $W = N \cdot hn$, impinges on a simple surface for 1 second, then the blanket impulse at a reflectivity *R* will look like:

$$P = \frac{N \cdot hv}{c}(1+R) = \frac{W}{c}(1+R)$$

Thus, electromagnetic theory and quantum theory give identical results. However it is essential different representations will be utillized in the electromagnetic and quantum theories of light. The concept of an electromagnetic wave of application to a light guesses existence of a wave in particular field(area) of space, which is capable to contain even some lengths of waves of light.

The light quantums are considered as a particle, which are localized in very small volume of space. The representation about a light wave guesses a continuous distribution of energy in a spatial wave, the representation about quantums guesses localization of energy in a small part of space. Both these incompatible representations reflect unity of a wave-corpuscle nature of light, which is shown at distribution of light and its interaction with substance.

PROPERTIES OF SUBSTANCE

De Broglie's Theory

The French physicist Louis Victor duc de Broglie has applied wave-corpuscle treatment and to particles of substance: as the light detects corpuscular properties, also particles of substance should detect undular properties. De Broglie has utillized already wave-corpuscle properties, accepted for a light, and their quantum rule and has expressed a wave length through the performances of a particle. For a light the corpuscular point of view determines energy and impulse accordingly by expressions

$$E = mc^2 \text{ and } p = mc.$$

The wave length of a particle of substance is determined under the formula:

$$\lambda = \frac{h}{P} = \frac{h}{mv}$$

At adding in a rule of quantization of a moment of momentum of an electron the impulse of an electron, will turn out a relation $nl = 2pr$. From it follows, that on length of a stationary orbit of an electron the integer of lengths of waves should be stacked.

Thus, the requirement of existence of allowed orbits, which was injected by Bohr frequency relation, acquires legible physical sense: the allowed orbits are orbits supposing formation on them of standing waves of an electron. For a rectilinear motion of particles de Broglie has offered expression, which is made by analogy with the equation for a spread flat light wave in the complex shape:

$$\psi = \psi_0 e^{i\left(\varpi t - \frac{x}{\lambda}\right)}$$

De Broglie's Waves

The specified waves have received a title of associated waves. Thus, the associated waves are traveling waves for freely propellent electrons or standing waves for electrons, which are bound in atoms. De Broglie's waves are not waves of propellent substance, they have not analog in classical physics.

De Broglie's guess about existence of undular properties of particles carries universal character: undular properties should have the electron, positive proton, neutron, atom, molecule, any moving object. However objects, which have major masses and move with usual velocities, will have a so small wave length To in comparison with the sizes of objects, that the phenomena of an interference and diffraction for them can completely be neglected.

The undular properties of a light are clearly shown in cases, when the wave length has compared to the sizes of bodies, with which the light interreacts. The lengths of waves of electrons in usual requirements have the order of the nuclear sizes, hence, the effects are characteristic for them which are observed usually for Roentgen rays. For such small lengths of waves it is possible to observe a diffraction on nuclear crystal lattices.

EXAMINATION OF PROPERTIES

The existence of undular properties of substance was revealed at observation of patterns of a diffraction of beams of particles. Legible patterns of an interference and the diffractions give bundles of electrons, which have almost identical impulses. These patterns are very similar to optical patterns of an interference and diffraction.

At change of quantity of a momentum of electrons and measuring of a wave length, it is possible to receive a relation between these quantities. It has appeared equal to a known relation

$$\lambda = \frac{h}{P}$$

This de Broglie's relation is valid for any kind of particles. The undular properties of particles were detected at observation of a pattern of the diffraction of the bundle of electrons on chips. At measuring a scattering angle it is possible to spot a wave length of electrons.

Thus, at known quantity of potential, which was utilised for acceleration of a bundle of electrons, their velocity is possible to calculate, and then to receive independent value of a wave length of an electron. The value of a wave length of an electron, which was spotted in such a way, completely coincides with quantity of a wave length, which is found from a diffraction pattern.

The undular properties of electrons also were detected in experience with a cathode rays at major velocity, which was passed through a thin metal foil on a photographic plate. After development of a photoplate on it the diffraction pattern from rings of different intensity was visible. However this diffraction pattern was stipulated not by Roentgen rays, which could arise at a concussion of high-velocity electrons with metal. It confirms presence of a stream of charged particles.

MECHANICS

Quantum Mechanics is the study of matter and energy on the subatomic scale. It was came into being because classical physics could not explain certain experimental results, such as why the photoelectric effect occured only when the light shone is above a certain frequency. Quantum Mechanics and Relativity became the foundation of modern physics.

Main Points

Principle of Quantatization

The fundamental principle of Quantum Mechanics, and its namesake, is that energy came in discrete particles called "quanta". This idea, the Quantatization of Energy, was put forth by Planck in his Quantum Hypothesis(1900), an explanation of blackbody radiation. It was a departure from Classical Physics, which asserted that energy could be infinately small.

Photoelectric Effect

This quantatization of energy approach was used by Einstein in his 1905 explanation of the photoelectric effect, the emission of electrons by a metal when light is shone on it. No electrons are emitted when the light is below a threshold frequency, no matter how intense the light. Einstein postulated that light, while exibiting wave properties as shone by the Double-slit Experiment, also came in the from of particles called "photons", discrete packets of energy.

The energy of a photon is proportional to the frequency of the light($E=hv$); thus, light with higher frequency had more energy and could excite the electrons, while light whose frequency were below the threshold could not, no matter how many photons there were.

Bohr's Model and Spectral Lines

Bohr also used the Quantatization of energy in his 1913 model of the atom. He stated that electrons revolved around the nucleus much like the planets revolve around the sun.

They could be found in predefined, stationary orbits, radiating no energy as they revolve. Only when they gain a quatum of energy equivalent to the difference between the energy of the orbits do they become excited and enter a higher energy orbit. This is consistent with Einstein's explanation of the photoelectric effect, and explains why spectral lines only occur at certain places.

Wave-particle Duality of Matter

In 1923 De Broglie expanded the Principle of Wave-particle Duality of Light put forth by Einstein to include all matter, called Principle of Wave-particle Duality of Matter. He asserted that all matter, not just light, exibit both wave-like and particle-like behaviour,

Quantum Matrix and Wave Mechanics

By now there was enough theories to put together a quantum picture of the atom. However, it was very difficult to sort through all the math. Heisenherg and Shroedinger simultaneously and independently worked to put forth a complete quantum mechanics.

Heisenberg employed use of matrices, and his model was called the matrix mechanics. His approach emphasized the quantum—discrete properties while Schroedinger concentrated on the wave properties. He tried to find a function that would encompass all informatin about a particle or system at any given time.

This is called the wave function. While debated ensued about which was correct, Schoedinger proved shortly after that the matrix mechanics and the wave function were mathematically equivalent.

Uncertainty Principle

Heisenberg then came up with the Uncertainty Principle, which stated that the position and velocity(momentum, which is mass×velocity) cannot be known precisely simultaneouely. The more accurately one is known, the less accurately the other can be measured.

The product of the uncertainty in position, and the uncertainty in momentum, is always greater than or equal to $h/(2\pi)$. While this is neglegible on the macroscopic level, subatomically, it causes great uncertainty. As a result, the world exists in statistical probabilities.

For example, a electron cannot be found precisely at one location. At any given chance, it has a certain probability of being found anywhere in the universe, with higher probabilities in regions around the atomic nucleus and less likelihood elsewhere. This can be calculated with the wave function.

Quantum Atomic Model

The quantum model of the atom calls for arbitrarily defined orbitals.

An orbital is a region around the nucleus where an electron has 90% chance of being found. Electrons have spin, a term used for the intrinsic angular momenta subatomic particles have. They have either a spin of +1/2 or –1/2. Two electrons in the same orbital of the same energy level must have opposite spin (Pauli Exclsion Principle).

Schroedinger's Cat

Since the wave function describes the world in terms of statistical probabilities and not certainties, a cat placed in an isolated system with equal chances of being found alive and dead at anytime *is* equally dead and alive at the same time.

When an observer is introduced to the isolated system, the wave function collapses and the cat is either dead or alive.

PRINCIPLES OF QUANTUM MECHANICS
UNCERTAINTY PRINCIPLE

One of the most important concepts of quantum mechanics is the uncertainty principle, which states that there is always some uncertainty when trying to measure the posistion and momentum of a particle. More specifically, the product of the uncertainty in position, Δx, and the uncertainty in momentum, Δmu, is always greater than or equal to h over 2π, where h is planck's constant.

When trying to determine the position of an electron, for example, we cannot know *exactly* where it is and where it's going at the same time. The more accurately one is measured, the less accurately the other is known. To we measure the position of the electron, for example, we can only obtain the information through the collision of the electron with another subatomic particle, such as a photon.

After the collision, the momentum of the photon changes—it "bounces" away, and from this change in momentum the position of the electron is calculated (this is how we see-photons bouncing off objects make possible vision).

However, the photon also gives the electron a change in momentum, so that now the position and velocity of the electron has changed from what it was. Thus, we can't know precisely the position and velocity simultaneously.

ELECTRON IN THE BOHR ATOM

In the Bohr model, electrons revolved around the nucleus in orbits. According to classical physics, they should radiate energy as they move and ultimately fall to the centre of the atom. Yet this is not so. The Quantum explanation is that electrons can only exist in predefined, discrete energy levels, and nowhere in between. The lowest energy level is not at the centre of the atom, and electrons may go no closer than the lowest level.

This explains why they don't collide into the nucleus. To move to another energy level, they must gain or lose *exactly* the energy difference between the levels. This explains spectral lines.

Another explanation is that, if electrons crashed into the nucleus, their velovity would be zero and the uncertainty would also be zero. This would be a direct violation of the uncertainty principle, a fundamental property of nature.

SPIN AND QUANTUM NUMBERS

It was discovered experimentally that subatomic particles have intrinsic angular momentum. That is, that angular momentum is not given by some interaction, but a property of the particles. This is a quantum property that classical physics has no explanation for.

The closest thing on the macroscopic level is rotation (since angular momentum leads to rotating), so this property is called "spin". Subatomic particles, however, don't really spin; they just have the angular momenta. This property is quantatized since the particles may only have certain discrete values of spin and not any value in between. Electrons have spins of either +1/2 or –1/2.

In the quantum model, electrons in the atom are given four quantum numbers to distinguish them. The first three distinguish the energy level, the shape of the orbital, and in which orbital the electron is located (has 90% chance of being found), and the last one gives the spin. The Pauli Exclusion Principle states that no 2 electrons in the atom can have the same 4 quantum numbers. If they occupy the same energy level and orbital, then they must have opposite spin.

NUCLEUS

The nucleus is a compact and dense region in the centre of the atom that makes up for most of its mass. It is made up of a group of subatomic particles known collectively as nucleons. The two states of nucleons are protons and neutrons.

Although the electron, and not the nucleus, is responsable for chemical reactions, the nucleus still holds much information about the atom. Since nucleons are about 2000 times heavier than electrons, the mass of the nucleus is approximately equal to the mass of the atom.

The number of protons in the nucleus determines the type of atom, and nuclear charge is the property by which the elements are arranged in the periodic table.

NUCLEAR ENERGY

A tremendous amount of energy is used to hold the nucleus together, since the repulsion between the positively charged protons is great. The bonding forces that hold the nucleus together are collectively known as the nucleur forces.

When heavy element nuclei are split into smaller nuclei, or light element nuclei combined to make one nucleus, tremendous energies are released. The former is known fission, and the latter, fusion, and they are the driving forces behind stars and nuclear weopons.

HISTORY

From Dalton's theory, it was thought that atoms were indivisible particles that made up everything in the world. When Rutherford made his gold foil experiment, however, he found that when high-velocity positive alpha particles bombarded a thing metal foil, while most passed straight through, some were reflected back or deflected.

This led him to postulate that most of the atom was empty space, with most of the mass concentrated in one dense region in the centre of the arom. Thus, most of the positive alpha particles passed through the empty space, but some hit the dense core and were reflected back by electric repulsion. He called this the nucleus of the atom.

Subsequently, improvements were made on the Rutherford model. The nucleu was theorized to contain both positive and neutral particles.

APPLICATION

The interactions in the nucleus, the strong and the weak forces, are the strongest forces in nature. Processes such as fission and fusion release much of this energy, and make possible new energy sources. In fact, fusion is the process by which stars generate light and heat, and thus the source from which all life on earth is derived. Other processes, such as radiation, have medical applications.

STRUCTURE OF NUCLEUS

The Dalton atomic model suspected no nucleus. It assumed that atoms were the fundamental building blocks of nature and no particles exist that are smaller than atoms. However, Rutherford's Gold-foil experiment led him to suspect that atoms had nuclei where most of its mass was located, and that the nucleus was positive. Bombarding Nitrogen with alpha particles, he noticed that the disintegration of the nucleus produced a positive particle equal to the mass of a hydrogen atom. He called this particle the proton and postulated that they made up the nucleus.

But a lot of positive charge concentrated in a small region would cause a lot of repulsion forces. So when Chadwick discovered the neutron in 1932, Heisenberg created the proton-neutron model of the nucleus. It also successfully explains isotopes.

THE PROTON

The proton is the nucleon that has a positive unit chage. Its mass is one atomic mass unit (amu, or 1.67×10^{-27} kg, 1,836 times the mass of an electron).

It is present in the nucleus of every element. Protons are responsible for the nuclear charge, the positive charge of the nucleus that deflected Rutherford's alpha particles. This charge also corresponds to the atomic number, which determines what kind of element the atom belongs to, and where on the periodic table the element belongs.

THE NEUTRON

THe neutron is an electrically neutral particle slightly heavier than the proton. They are present in numbers greater than or equal to the number of protons in every element except Hydrogen (whose nucleus is a single proton), and variations in the number of neutrons of elements accound for isotopes.

The neutron has a maganetic moment, which suggests that its internal structure is made up of electrically charged components whose net charge is zero. The neutron is actually the fusion of a proton and an electron (since in beta decay a neutron is turned into a proton and an electron) and a neutrino.

ISOTOPES

Isotopes are atoms of the same element that have differenet mass. This is explained by the proton-neutron model of the nucleus. The number of protons account for which element it is and thus how it behaves, but the number of neutrons there are determines which isotope it is, and there can be more neutrons than there are protons.

For example, elemental Hydrogen has one proton (H-1) and a mass of one amu. But an isotope of Hydrogen, deuterium, has a proton and a neutron, and a mass of two amu. As another example, Carbon-12, Carbon-13 and Carbon-14 all have 6 protons (this determines that it is Carbon atom) but have 6, 7 and 8 neutrons, respectively.

Sometimes there are too many neutrons in the nucleus. When this happens, the neutron goes under beta decay, a type of radioactive decay. The neutron splits into a proton, an electron (beta particle) and a neutrino. The proton stays in the nucleus, but the beta particle and the neutrino(an electrically neutral, light or massless particle) are emitted, and energy is released, carried by the electron (when the electron doesn't carry all the energy, the rest is carried by the neutrino).

CLASSIFICATION

Nucleons are generally knoen as a baryon. Baryons are elementary particles consisting of three quarks and no antiquarks. They obey the Fermi-Dirac statistics, which encompass all elementary particles that obey the Pauli Exclusion Principle. The proton is the lightest baryon and is made of two up-quarks and one down-quark. The neutron is made of two down-quarks and one up-quark.

Nucleon's antiparticles are the antiproton and the antineutron, which have the same mass as the corresponding nucleons but opposite electric charge and magnetic moment.

TYPES OF RADIOACTIVITY

The study of radioactive material started with Becqueral in 1896, when he discovered that some of his films became unuseable, as if they had been exposed to light, even though they were kept in a sealed opaque container. Further investigation revealed a substance that gave a green glow in the dark. This was uranium, and Becqueral discovered the X-ray. Marie Curie then studied radioactive material, discovering several radioactive substances. This research led to a better understanding of radiation, and peopls started bombarding elements with particles emitted from radiation.

Rutherford's experimentations, for example, led him to his atomic model. These bombardments also led to the discovery of many particles, such as the proton. The nature of radiation was also better understood as the identity of the radiated particles came to be known, and theories of radioactive decay were devised.

These experiments did not reveal a complete picture of radiation, until Fermi's systematic bombardment of all elements. The results led him to envision a chain reaction, where the bombardment of an atom led to its splitting and emitting more particles that bombarded the other atoms. This was carried out in 1942, and led to the development of the A-bomb shortly after.

ALPHA PARTICLES

Alpha radiation (or alpha rays) was distinguished and named by E. R. Rutherford in 1909, who found by measuring the charge and mass of alpha particles that they are the nuclei of ordinary helium atoms. Alpha particles consist of two protons and two neutrons.

BETA PARTICLES

Beta radiation (or beta rays) was identified and named by E. Rutherford too. He found that beta rays consists of high-speed electrons. Unlike alpha and gamma particles, whose energy can be explained as the difference of the energies of the radioactive nucleus before and after emission, beta particles emerge with a variable energy.

This apparent violation of the law of conservation of energy led to the hypothesis that a second undetected particle, the neutrino, is emitted along with the electron and shares the total available energy.

In some forms of induced, or artificial, radioactivity, the electron's antiparticle, the positron, is emitted from the excited nucleus; the positron in this case is also called a beta particle and denoted by b^+ (the ordinary beta particle is b^-).

GAMMA RADIATION

Gamma radiation is high-energy photons emitted as one of the three types of radiation resulting from natural radioactivity. It is the most energetic form of electromagnetic radiation, with a very short wavelength (high frequency). Wavelengths of the longest gamma radiation are less than 10^{-10} m, with frequencies greater than 10^{18} hertz (cycles per sec). Gamma rays are essentially very energetic X rays; the distinction between the two is not based on their intrinsic nature but rather on their origins.

X rays are emitted during atomic processes involving energetic electrons. Gamma radiation is emitted by excited nuclei or other processes involving subatomic particles; it often accompanies alpha or beta radiation, as a nucleus emitting those particles may be left in an excited (higher-energy) state.

The applications of gamma radiation are much the same as those of X rays, both in medicine and in industry. In medicine, gamma ray sources are used for cancer treatment and for diagnostic purposes.

Some gamma-emitting radioisotopes are also used as tracers. In industry, principal applications include inspection of castings and welds. Data from artificial satellites and high-altitude balloons have indicated that a flux of gamma radiation is reaching the earth from outer space, thus opening up the field of research known as gamma-ray astronomy.

DECAY ALPHA DECAY

There are three types of radioactive decay, named by the type of particle they produced. All radioactive decay is the result of an unstable nucleus and release energy. The first is the alpha decay, which emits an alpha particle, a positively charged and heavy particle that does not penetrate too well.

It was shown that the alpha particle is actually a Helium nucleus, consisting of two protons and two neutrons. When the unstable nucleus emits the alpha particle, the atomic number of the atom decreases by two and the atomic mass number decreases by four.

BETA DECAY

The beta particle is an electron. When beta decay occurs, a neutron in the nucleus converts into a proton and an electron. The electron is emitted and takes the released energy with it. When it does not carry enough enery, however, the remaining is carried by a small, mass less (or very light) neutral particle called the neutrino. The beta particle is therefore negative and more penetrating than the alpha particle.

GAMMA DECAY

The Gamma radiation is unaffected by magnetic field as it travels; therefore it is electrically neutral. It is actually a type of electromagnetic radiation, and not a particle of significant mass. Gamma rays is one of the most energetic types of electromagnetic radiation, because its wavelength is very short and its frequency very high, (they are therefore invisible to the eye).

Gamma radiation often accompanies alpha decay or beta decay, since the nucleus emitting those particles may still be excited, and the gamma ray is the result of the extra energy. Gamma rays are the most penetrating type of radiation.

NATURAL GROWTH

Radioactive decay obeys the rules of natural (exponential) growth or decay. That is, the rate of decay is directly proportional to the amount of undecayed substance present. (Chemically, the rate of reaction is directly proportional to the amount of unreacted reactant.)

Thus, the first derivative of the substance, say, Uranuim (U) with respect to time is proportional to the amount of undecayed U present.

$d[U]/dt = k[U]$.

Solving this gives $d[U]/[U] = kdt$, or

$\ln[U] + C = kt$, $[U] = ae(kt)$,

where a is calculated to be the amounf or Uranium at the start of the decay.

A proporty of the exponential graph is that starting at any arbitrarily chosen time, the time it takes for the amount of undecayed Uranium to decrease by half (or any fraction) is the same. This period of time, the time it take for half the sample to undergo radioactive decay, is called the Half-life of that element.

After one half-life, the amount of unreacted sample left is alway one half the amount of unreacted sample before that period of time. Thus, if the half life of a particular sample is 10 years, then after 20 years one-quarter of undecayed sample will be left.

INDUCED RADIOACTIVITY

The induced radioactivity was discovered by French physicists Iren Joliot-Curie) and Frederic Joliot-Curie. The first synthetic nuclear reaction was realized by Rutherford. It is possible to register the charged particles which are taking off at a radioactive decay. It allows to use atoms of the synthetic radioactive isotopes.

NUCLEAR REACTIONS

Nuclear reactions. There are 2 stages of nuclear reactions. The first stage of nuclear reaction consists in entrapment by a kern of a particle. For a small time interval the energy is uniformly proportioned between all particles of a compound nucleus.

The second stage of nuclear response (start of a particle from a compound nucleus) occurs as a result of casual diversions from a uniform distribution of energy. There is a time between entrapment of particles and their start from a kern, which is significant more then nuclear time $t_n = 10^{-22}$ s. All process of nuclear reaction is possible to figure by the formula:

$$z_1X^{Al} + a \rightarrow z_2Y^{A2} \rightarrow z_3C^{A3} + b$$

$$z_1X^{A1} - \text{intial kern} : a, b - \text{particles}:$$

$$z_2X^{A2} - \text{componund nucleus} : z_3X^{A3} - \text{terminating kern}$$

DIVISION OF KERNS

The heavy kerns are labile. The kern, which was exited as a result of a neutron capture, can be divided into identical parts. The nuclear fission on two debrises is accompanied by eduction of huge energy. The debrises, which are formed at division, are radioactive and can emit neutrons.

As the result the exuberant number of neutrons is formed, that gives in radio-activity of yields of decay.

CHAIN REACTION OF DIVISION

The division of kerns of uranium occurs at interaction to prompt or sluggish neutrons, but the irradiation by sluggish neutrons is more effective. The practical interest is represented by reactions, at which the major energy is oozed and there are neutrons, which give in magnification of number of the acts of division. For realization of chain reaction the maintenance of fixed number of neutrons is necessary.

For an estimation of number of neutrons is inlet coefficient of manifolding of neutrons k. It determines the ratio of number of neutrons, which arise at reaction, to number of neutrons, which participated in reaction. If $k >= 1$, the reaction will be explicated.

The minimum sizes of a fissile region, at which the realization of a chain reaction is possible, term as the critical sizes. The minimum quantity of a fissionable material necessary to sustain a nuclear chain reaction is called the critical mass.

CONTROLLABLE REACTION OF DIVISION

The controllable reactions occurs in a spacial nuclear reactors. Reactors can be used for research or for power production. A research reactor is designed to produce various beams of radiation for experimental application; the heat produced is a waste product and is dissipated as efficiently as possible.

In a power reactor the heat produced is of primary importance for use in driving conventional heat engines; the beams of radiation are controlled by shielding.

A-BOMB

Practical fissionable nuclei for atomic bombs are the isotopes uranium-235 and plutonium-239, which are capable of undergoing chain reaction. If the mass of the fissionable material exceeds the critical mass (a few pounds), the chain reaction multiplies rapidly into an uncontrollable release of energy.

An atomic bomb is detonated by bringing together very rapidly (e.g., by means of a chemical explosive) two sub critical masses of fissionable material, the combined mass exceeding the critical mass.

An atomic bomb explosion produces, in addition to the shock wave accompanying any explosion, intense neutron and gamma radiation, both of which are very damaging to living tissue. The neighborhood of the explosion becomes contaminated with radioactive fission products. Some radioactive products are borne into the upper atmosphere as dust or gas and may subsequently be deposited partially decayed as radioactive fallout far from the site of the explosion.

ELEMENTARY PARTICLES

All subatomic particles, such as nucleons or electrons, and the particles they release when undergoing processes such as radioactive decay, are organized into classes.

All particles that obey the Pauli Exclusion Principle (no two particles can have the same four quantum numbers in a system, such as an atom) are called fermions because the Fermi-Dirac statistics apply to them. All particles that don't obey the principle are called bosons because the Bose-Einstein statistics apply to them.

FERMIONS

Fermions have half-integral spin (intrinsic angular momentum). They are deemed anitsocial particles because they obey the exculsion principle, meaning that no two fermion with the same spin can be found in a system. The antiparticles for fermions have the same mass but opposite charge and magnetic moment as the corresponding particle.

BARYONS

The first type of fermion is the baryon. They are those particles that experience the strong nuclear force. They consist of three quarks (quark triplet). The most notable baryons are the nucleons, the two states of which are protons and neutrons, the particles that make up the nucleus. Other, heavier baryons, called hyperons also exist, but they are unstable, and usually decays to a form of nucleon(heavier baryons include the sigma, the lamda and the delta particles). Since all baryons decay to the nucleon form, the nucleons are known as the baryon ground state.

All baryons obey the law of baryon family numbers (the family number is +1 for particles such as protons and -1 for antiparticles such as antiprotons), which states that the sum of the family number of all reactants in an interaction is equal to the sum of the family number of all products.

This means that, for every particle created or destroyed, a corresponding antiparticle is also created or destroyed. For example, a proton colliding with an antiproton will result in the mutual destruction of both particles, or, a proton and an antiproton can be created simultaneously.

LEPTONS

Leptons are fermions that do not experience the strong nuclear force. They are only weakly interactive, and can be created by nuclear decay. Leptons include such particles as electrons, muons, neutrinos and the tau particle.

So, for example, an electron and a neutrino can be created by the decay of an unstable neutron to a proton. As fermions they obey the exclusion principle, so no two electrons can occupy the same energy level and orbital of an atom (a system) and have the same spin. Leptons are the lightest fermions. There are three separate laws concerning the conservation of lepton family numbers (the family numbers for leptons, like for baryons, is +1 for particles and –1 for antiparticles), for the electrons, muons and taus.

BOSONS

Bosons are particles that, unlike the fermions, do not obey the exclusion principle. Generally, fermions are particles that compose atomic structure but bosons carry out the interaction (forces) between fermions. Bosons have integral spin and are deemed social because many bosons can by placed in the same system in the same energy state at once.

MESONS

Mesons are a class of bosons whose mass falls between the lighter lepton fermions and the heavier baryon fermions. They consist of two quarks (quark pair), and include such particles as the p meson (pion) and the *K* meson (kaon). It was previously thought that the muon was a type of meson (m meson) because of its large mass (about 200 times that of the electron), but the behaviour of the muon doesn't fit into the description of mesons, so it was reclassified.

The pion is accountable for carrying out the strong nuclear force, which is the force that holdes the nucleus together, despite tmememndous electric repulsion.

OTHER BOSONS

Other bosons include the W and Z particles are responsible for carrying out the weak nuclear force and wjocj are about 200 heavier than the proton, and the photon, the particle form of light. The photon is its own antiparticple, as is the neutral pion.

Chapter 7

Chemical Bonding

The electronic configurations of the elements, apply in principle only to isolated atoms – atoms separated by distances over which no mutual interactions of their electronic orbitals can occur (infinite distance).

This condition is never met in condensed phases (i.e., liquids and solids); it is only encountered in high vacua where atoms move over long distances without mutual interaction. Under normal conditions, particularly in the mentioned condensed phases, atoms are separated over distances controlled, in essence, by the dimension of their respective outermost occupied electronic orbitals.

Whenever the outer electron shells of two or more atoms come in contact with each other (overlap to any extent), the potential for interaction (reaction) exists. Spontaneous reaction (consisting of a rearrangement of the electronic orbitals and/or actual transfer of electrons from one atom to another) will take place whenever such a rearrangement results in a lower energy configuration.

This means that the driving force for mutual interaction and rearrangement of electronic configurations is in most instances (but not always) manifested through the release of heat (a form of energy) to the environment. [Typical case: Mg + 1/2 O_2 σReaction Product (MgO) + (light + heat) i.e. energy.]

Since the same amount of energy must be supplied to the system if the original state – i.e., separation of the species – is to be reestablished, the reaction partners are "bonded together" by comparable energies. The "strength" of the bonds can obviously vary from system to system with the nature of the electronic rearrangement.

Even the inert gases, particularly the heavier ones like xenon, are capable of forming associations with other atoms. Sometimes we find that two atoms assume a more stable state by sharing electrons; at other times, an atom may transfer electrons to another atom in order to achieve a greater stability.

In still other instances, the rearrangement may simply be an orbital distortion or an internal charge redistribution. In either event the mutual benefit that accrues is the formation of what is commonly called a chemical bond. Through these bonds atoms combine with each other to form very different kinds of particles referred to as molecules and ions.

NATURE OF CHEMICAL BONDS

In order for a chemical bond to be formed between two atoms, there must be a net decrease in the energy of the system (the two atoms): the ions or molecules produced by electronic rearrangements must be in a lower energy state than the atoms were prior to interaction, prior to bond formation. Since atoms of each of the elements have different electronic structures, the variety of possible chemical bonds (differing from each other in at least some small way) is considerable and is even further increased by the effects of neighboring atoms on the bond under consideration.

The modes of bond formation can be categorized into two basic types, each representing a type of bonding. The bonding types are called electrovalent (or ionic) bonding and covalent bonding. Electrovalent bonding arises from complete transfer of one or more electrons from one atom to another; covalent bonding arises from the sharing of two or more electrons between atoms. Since these models represent the limiting cases, we can anticipate that most real bonds will fall between these two extremes. (Two additional types of bonding, metallic bonding and Van der Waals bonding, will be discussed later.)

Before discussing these models in detail it is appropriate to consider the relationships between the electronic structures of atoms and their chemical reactivity. The inert gases (Group VIII) are the most stable elements with regard to bond formation, i.e. toward electronic rearrangements. It is therefore useful to examine the reasons for their stability. Inert gases all have electronic structures consisting of filled subshells. For all but helium the outer (or valence) shell contains eight electrons, with filled s and p sublevels (ns^2p^6). The electronic structure of helium is $1s^2$, which is equivalent to the structure of the other inert gases since there is no 1p sublevel. Inert gases have high ionization energies because each electron in the sublevel of highest energy is poorly screened from the nucleus by other electrons in its same sublevel.

Each electron "sees" relatively high positive charge on the nucleus, and a large amount of energy is required to remove it from the atom. Inert gases have very low electron affinity because any added electron must enter a significantly higher energy level. We find, therefore, that the electronic structures of inert gases are particularly resistant to changes by either loss or gain of electrons and, further, that atoms of other elements with fewer or more electrons than inert gas configuration tend to gain or lose electrons, respectively, to achieve such inert gas structure.

ELECTROVALENT (IONIC) BONDING

An electrovalent bond is formed by the transfer of one or more electrons from one atom to another. Consider first atoms that have electronic structures differing from an inert gas structure by only a few, (1, 2 or 3) electrons. These include the representative elements of Groups I, II and III in the Periodic Table, which have respectively 1, 2 and 3 electrons more than a neighboring inert gas, and the representative elements of Groups V, VI and VII, which have respectively 3, 2 and 1 electrons less than a neighboring inert gas.

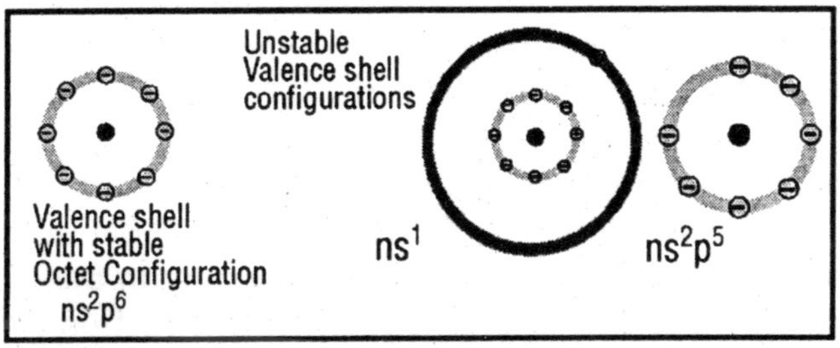

Fig. Stable and Unstable Valence shell Configigutations.

The elements of Groups I, II and III can form the electronic structure of an inert gas by losing their outer 1, 2 and 3 (valence) electrons. (The resulting species are positively charged ions.) In a similar electron transfer which, however, involves the acquisition of electrons in the outer valence levels, elements of Groups V, VI and VII form an inert gas electronic structure (by formation of negatively charged ions). It is through the electron transfer between an electron-losing element and an electron-gaining element that compounds are formed which involve electrostatic attraction (electrovalent bonds) of oppositely charged species called ions.

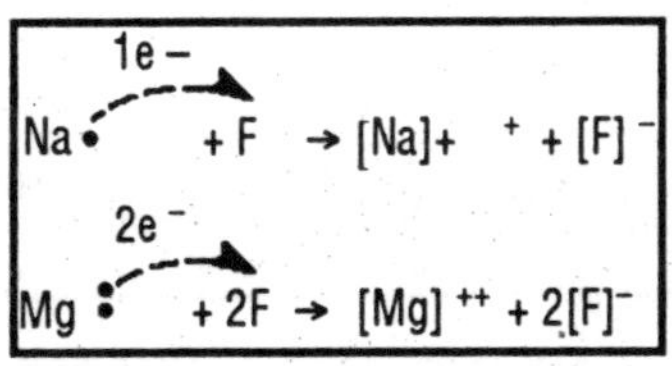

Fig. Electrostatic Attraction of Oppositely Charged Species.

(In the notation [Na], the dot indicates the outermost electron which is in excess of the rare gas configuration. It is referred to as a valence electron.) Elements immediately following the inert gases (in the horizontal columns of the Periodic Table) lose electrons, and those immediately preceding the inert gases gain electrons on interaction. The resulting compounds are called electrovalent; the valence number (charge on the ion) of a particular element when it forms an electrovalent compound is given by the number of electrons lost or gained in changing from the atomic to the ionic state. The stoichiometric formula of an electrovalent compound reflects the ratio (usually very simple) of positive to negative ions that gives a neutral aggregate. Hence, the ions Na+ and F– form a compound whose formula is NaF because these ions are singly charged and are present in the compound in a one-to-one ratio. Magnesium nitride, composed of Mg_2^+ and N_3^-, has the formula Mg_3N_2 because this composition represents electroneutrality.

The electrovalent bond is the result of electrostatic attraction between ions of opposite charge. This attractive force accounts for the stability of these compounds, typified by NaF, LiCl, CaO, and KCl. The ions individually possess the electronic structures of neighboring inert gases; their residual charge arises from an imbalance in the number of electrons and protons in their structures. Isolated ions and simple isolated pairs of ions, as represented by the formula NaCl, exist only in the gaseous state.

Their electrostatic forces are active in all directions; they attract oppositely charged species and thus can form regular arrays, resulting in ordered lattice structures, i.e. the solid state. Even in the liquid state and in solutions (where disruptive thermal forces reach values close to that of the attractive electrostatic bonding forces) attraction between ions and with other species remains effective.

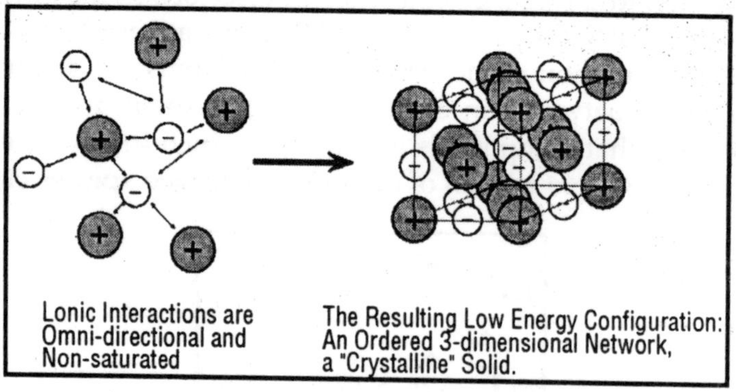

Fig. Formation of a Crystalline, Ordered Body (a SOLID), a Direct. Consequence of Ion Interaction in Conjunction with Energy Minimization..

ENERGETICS OF IONIC BONDING

Ionic bonding is the simplest type of chemical bonding to visualize, since it is totally (or almost totally) electrostatic in nature. The principle of the energetics of ionic bond formation is realized when considering the formation of NaCl (our common salt) from its constituents Na (metal) and Cl_2 (chlorine gas). Formally, this reaction is:

$$Na(s) + 1/2(g) \rightarrow NaCI(S)$$

$$\Delta H = -414 kJ/mol$$

The equation as written indicates that 1 mole sodium reacts with 1/2 mole chlorine (Cl_2) under formation of 1 mole (ionically bonded) sodium chloride; this reaction is accompanied by the release (–) of 414 kJ of energy (Δ H), referred to as the heat of reaction. From earlier considerations it is clear that electronic rearrangement (reaction or bond formation) takes place because the resulting solid product (NaCl) is at a lower energy state than the sum total of the energies of the original components.

The energetics associated with ionic bond formation may be determined quantitatively by considering the energy changes associated with the individual steps leading from the starting materials to the final product (Haber-Born cycle).

The bond formation in NaCl may be formally presented as an electron-transfer reaction:

Ionizaton of Na:

$Na(gas) \rightarrow Na^{+} + Ie$ (E.I. = +497 kJ/mole)

The energy change associated with this step, energy of ionization (E.I.),is

+497 kJ/mole.

- *Acquisition of one electron by Cl:*

$Cl (gas) + 1e \rightarrow Cl -$ (E.A. = –364 kJ/mole)

The energy change associated with this step, electron affinity (E.A.), is –364 kJ/mole. The minus sign reflects an energy release, a lowering of the energy state associated with the achievement of stable rare gas configuration by chlorine.

So far, the energy balance appears positive (Δ E = +133 kJ); this means the reaction is not favored since the final products are at a higher energy state than the starting products. However, there are additional steps involved since:

- *Vaporization of Na:*

Na (metal) → Na (gas) (ΔH_V = +109 kJ/mole)

The energy required to transform Na (metal) into Na (gas), the latent heat of vaporization (ΔH_V), is 109 kJ/mole (now reaction appears even less favorable).

Dissociation of Cl_2:

$Cl_2 \rightarrow 2Cl$ (E.D. = +242 kJ)

The energy associated with breaking up the (stable) chlorine molecule into two reactive chlorine atoms, dissociation energy (E.D.), is +242 kJ/mole. (Since the formation of one mole NaCl involves only 1 mole Cl, and since 2 moles are formed from one mole Cl_2, the energy required in this step is 1/2 E.D., or 121 kJ.)

The total energy change associated with reactions (1) through (4), Δ HT = +364 kJ This indicates that one or more basic reactions which lead to an overall decrease in energy are still unconsidered.

Unconsidered as yet is the coulombic attraction of the reaction products which are of opposite charge and an energy term associated with the formation of the "solid state" product, NaCl.

The two ionic species, originally at zero energy (infinite distance of separation), attract each other with an accompanying energy change (decrease).

The energetics associated with the approach of the reaction partners is best considered by fixing the position of one, for example Na+, and letting the other (Cl^-) approach: as Cl^- approaches Na^+, E decreases according to Coulomb's law with $e^2/4\pi\varepsilon_o$ r. With the approach of the two oppositely charged ions, the outermost electronic shells will come into contact and a repulsive force will become active as the shells interpenetrate.

The repulsive force (Erep) increases with +b/r12 and thus is only active in the immediate vicinity of the sodium ion, but at that stage increases rapidly.

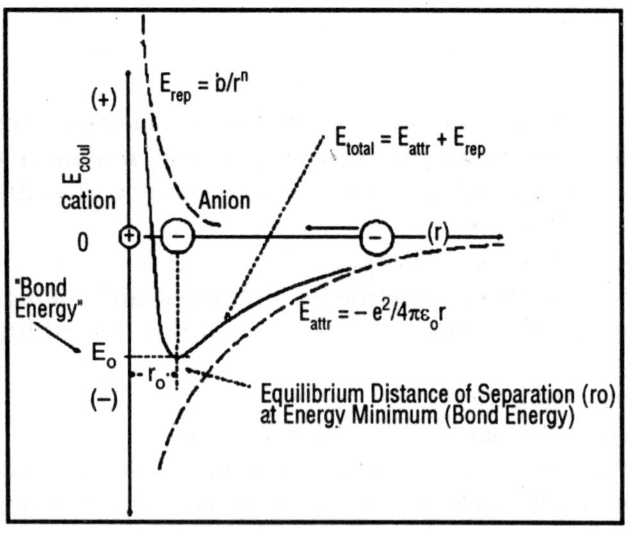

Fig. Energetics of Ionic Bonding.

The two ions at close proximity are under the influence of both attractive and repulsive forces and will assume a distance of separation at which the two forces balance each other - a distance which is referred to as the equilibrium separation (ro) and which corresponds to he energy minimum for the NaCl molecules:

$$E_{coul} = -\frac{e^2}{4\pi\varepsilon_o r_o} + \frac{b}{r_0^{12}}$$

where e = electronic charge, ε_o = permittivity of free space (8.85 x 10^{-12} F/m) and (b) is a constant. (ro for gaseous molecules may be obtained from physical measurements.)

If the presently considered coulombic (attractive) energy term is taken into consideration as reaction , it will decrease the overall positive energy change, ΔH_T; it will, however, not make it change sign from (+) to (–). Thus, NaCl molecules in gaseous form are not the final reaction product; nor would reaction occur if ΔH_T remained positive.

Considering an ionically bonded, gaseous Na^+Cl^- molecule, it is clear that its electrostatic forces (+) and (–) are not saturated - they remain active in all possible directions. This means that Cl^- will attract Na^+ ions from other directions as will the Na^+ atoms attract additional Cl^- ions. The result of these attractive forces in ali directions is the formation of a "giant size" ionic body - a "solid" body of macroscopic dimensions.

From the preceding it is recognized that the minimum energy configuration is given by a body in which Na^+ and Cl^- are arranged with extreme periodicity and order since any ion located outside of its "equilibrium position" will be in a "higher energy configuration"; such an ordered body is referred to as a crystalline solid, or frequently just called a solid.

The total energy change associated with the formation of one mole of crystalline ionic solid from its ionic constituents is given as:

$$E_{cryst} = -\frac{MN_A e^2(Q_1Q_2)}{4\pi\varepsilon_o r_o}\left(1 - \frac{1}{n}\right)$$

where M = 1.747 ("Madelung" constant for NaCl, reflecting multiple interactions for the particular geometric arrangement of ions in the solid), NA = Avogadro's number, Q = number of charges per ion (1 for Na^+ and Cl^-), ro= equilibrium distance of separation of ions, and n = repulsive exponent (n = 12 for NaCl). [The relationship for the crystal energy (Δ Ecryst) is readily obtained from equilibrium energy considerations:

$$E_{Coul} = -\frac{e^2}{4\pi\varepsilon_o r} + \frac{b}{4\pi\varepsilon_o r^n}$$

At equilibrium distance of ion separation (ro), (dE/dr) = 0.

Thus:

$$\left(\frac{dE}{dr}\right)_{r_o} = \frac{e^2}{4\pi\varepsilon_o r_0^2} - \frac{nb}{4\pi\varepsilon_o r_o^{n+1}} = 0$$

And

$$b = \frac{e^2 r_o^{n-1}}{n}$$

And

$$E_{o(Coul)} = -\frac{e^2}{4\pi\varepsilon_o r_0} + \frac{e^2}{n4\pi\varepsilon_o r_o}$$

$$E_{o(Coul)} = -\frac{e^2}{4\pi\varepsilon_o r_o}\left(1 - \frac{1}{n}\right)$$

For a "molar" crystal with ionic charges Q_1 and Q_2, the molar Δ Eo(Coul) is thus given as:

$$\Delta E_{cryst} = \frac{MN_A Q_1 Q_2 e^2}{4\pi\varepsilon_o r_0}\left(1 - \frac{1}{n}\right)$$

Here M and NA stand, as indicated above, for the conventional terms, the Madelung constant and Avogadro's number.]

For the present system (NaCl):

$$\Delta E_{cryst} - 777 kJ/mole.$$

Now considering reactions:

$$\Delta E_{Total} - 414 kJ/mole$$

which is identical with the value experimentally determined and given for the reaction:

$$Na(metal) + CI_2(gas) \rightarrow NaCI(Solid)$$

$$\Delta H_{Reaction} = -414 kJ/mole$$

In the Haber-Born cycle, the reaction energy (Δ H) associated with the formation of NaCl from Na + Cl2 may be summarized as:

$$\Delta H = E.I. + E.A. + \Delta H_V + 1/2 E.D. + \Delta H_{Cryst}$$

Δ H, the heat of reaction, may thus be obtained from the energetics of the steps leading to the end product. In most instances, however, the reaction energy (Δ H) is determined experimentally in a calorimeter and the Haber-Born cycle is used to obtain the value of Δ Ecryst or E.A., which are both extremely hard to come by.

Conclusions: A primary drive for atomic interactions leading to "bonding" is the achievement of valence shell octets which exhibit a high degree of stability. If atoms on the left and right side of the Periodic Table interact, i.e. atoms with a large difference in electron affinity (Δ E.A.), stabilization is achieved by electron octet formation through charge transfer.

The reaction products exhibit opposite charges (⊕ cations, ⊖ anions) and are subject to Coulombic attraction → ionic bonds are formed. Since the electrostatic forces are non-directional and non-saturated, energy minimization will result in the formation of macroscopic bodies that are highly ordered on the atomic scale, crystalline ionic solids. Ionic solids have mostly predictable, basic properties:

- Atomic arrangements are a function of the ion size ratio, the charge ratio of ions, and their electronic structure;
- Electrical and thermal conductivities are expected to be low because the high stability of the octets formed results in bound electrons which do not contribute to conduction (there is a large "energy gap" to be crossed for electrons to move into a higher energy state);
- Optically, ionic solids are mostly transparent, or translucent, reflecting octet stability of the electrons and macro- or micro-crystallinity;
- Elting points are high, increasing with the electronic charges on the cations and anions;
- Ionic solids are normally hard and brittle.

COVALENT BONDING

Wave Mechanical Concepts and Conclusions

In 1924 L. DeBroglie advanced the hypothesis that all matter in motion possesses wave properties and can be attributed a particle wavelength

$$\lambda p = h/mv$$

where h is the Planck constant, m is the mass of the matter and v is its velocity.

The credibility of this hypothesis was established by Davisson and Germer in 1927 when they demonstrated that electrons (like electromagnetic radiation) are diffracted by crystal lattices. An important consequence of the dual nature of matter (it exhibits both particle and wave properties) is the uncertainty principle established in 1927 by W. Heisenberg. It states that it is impossible to simultaneously know with certainty both the momentum and position of a moving particle:

$$(\Delta p_x)(\Delta x) \geq h$$

We can paraphrase the uncertainty principle in the following manner: If the energy of a particle is known (measured) with high precision, its location is associated with a high degree of uncertainty.

If electrons occupied simple orbits (as postulated by Bohr-Sommerfeld), their momentum and position could be determined exactly at any moment - in violation of the uncertainty principle. According to Heisenberg, if the energy of an electron is specified with precision (sharpness of spectral lines), its location can only be specified in terms of the probability of finding this electron in a certain location (volume element).

These arguments give rise to the concepts of probability density and electron cloud which are inherent to the wave-mechanical electron concept emanating from the solutions of Schrödinger's wave equation which relates the energy of an orbiting electron to its wave properties. When solving exactly the Schrödinger wave equation for an electron in a hydrogen atom, a quantization results according to which electrons can only assume certain energies which are in quantitative agreement with those obtained from the Bohr theory.

In comparison to the Bohr theory, significant differences are observed for the orbital quantization (orbital quantum number l) which specifies the orbital shape. For n = 1, l = 0 (1s orbital), wave mechanics predicts a spherical electron density distribution with a maximum probability density (ψ^2) at a distance of 0.529 Å (a_o) from the nucleus. For l = 1 (p orbitals), however, it is found that the orbitals (three) form lobes aligned with rectangular coordinates. For l = 2 (d orbitals), five complex orbital configurations are obtained.

Previous considerations suggest that all elements attempt to assume a stable octet configuration with eight electrons in the valence shell. (For hydrogen, with only one shell occupied, the stable configuration consists of two electrons in the K shell - which is the maximum number of electrons that can be accommodated.) Stable octet formation is possible through electron transfer (and ionic bond formation) when, for example, elements in columns IA, IIA and IIIA react with elements in columns VA, VIA and VIIA, respectively.

In these instances, the reaction partners exhibit rather pronounced differences in electron affinity and upon reaction one assumes octet configuration by losing one or more electrons while the other does so by acquiring the missing number of electrons.

This mechanism is clearly not possible if H reacts with H to form an H_2 molecule where two atoms are bonded together. The same argument holds for the formation of N_2, Cl_2 and O_2 molecules. Inert gas configuration (octet configuration) in such elements is achieved by a mechanism called orbital sharing and the resulting bond is called covalent, or electron–pair bond.

A covalent bond is somewhat more difficult to visualize than an ionic or electrovalent bond because it involves the sharing of a pair of electrons between atoms. The stability of this bond can be attributed to the complex mutual attraction of two positively charged nuclei by the shared pair of electrons.

In principle, the bond can be understood if it is recognized that both electrons in the bonding orbital spend more time between the two nuclei than around them and thus must exercise attractive forces which constitute the bond. In this arrangement it is clear that each electron, regardless of its source, exerts an attractive force on each of the "bonded" nuclei.

The pair of electrons in a covalent bond is unique to the extent that the Pauli exclusion principle precludes the presence of additional electrons in the same orbital. Furthermore, the pairing phenomenon neutralizes the separate electronic spins of the single electrons, and the resulting electron pair with its zero spin momentum interacts less strongly with its surroundings than do two independent electrons. [Covalent bonds are conveniently symbolized through the dot notation, introduced first by Lewis]

LEWIS NOTATION

In LEWIS notation the covalent molecular bond is indicated as a BAR or as two DOTS (standing for the paired electrons)

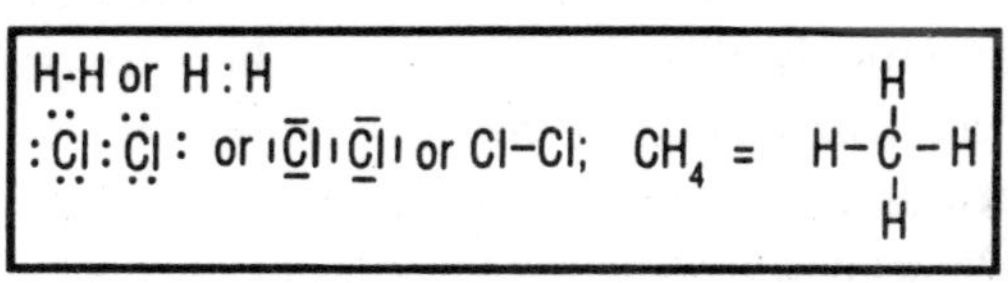

Fig. Lewis Notation.

Formal valence shell octet stabilization can be achieved by electron sharing, whereby one electron from each reaction partner share - spin paired - the molecular bonding orbital.

Quantum mechanics makes it possible to rigorously describe these bonds for very simple cases such as the hydrogen molecule, which is composed of two protons and two electrons. It can thus be shown that the potential energy for the system reaches a minimum for a certain equilibrium distance between the nuclei, with increased electron density between the nuclei. At shorter distances between the nuclei repulsive forces are found to increase very rapidly.

In the hydrogen molecule (H_2), the two hydrogen atoms are effectively linked together by one molecular electron orbital, termed a σ orbital, which comprises both atoms and contains two electrons. Each of these two electrons can be considered to have originated from one of the two atoms - they were originally both 1s electrons with the same spin value (s = +1/2). In the molecular orbit comprising both atoms, the spins of

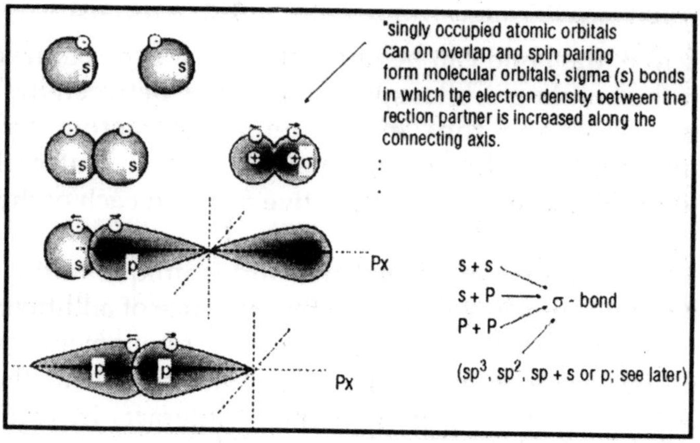

Fig. Covalent Bond Formation.

The two electrons must align anti-parallel (opposite spin). This spin-pairing process results in a considerable release of energy and thus contributes significantly to the stability (strength) of the covalent bond formed.

It is interesting that according to Newtonian mechanics, no stable configuration can arise from the placement of two electrons into the same region (orbit) - wave mechanics, however, predicts increased stability from such configurations. (Similar spin-pairing occurs in the filling of atomic orbitals.) In the molecular case, spin-pairing has the consequence that both the probability distributions (ψ^2) and spatial distribution of electrons are such that maximum overlap of orbitals of combining atoms occurs.

The bond formed in hydrogen is a single bond, containing two electrons with paired spins. This formulation predicts that there will be a fixed internuclear separation which, for the hydrogen molecule, has the value of 0.74 Å. [σ bonds are formed not only by s orbital overlap, but also by p-s and p-p orbital overlap as well as by the overlap of s or p orbitals with hybridized orbitals.]

It is important to recognize that octet stabilization by electron orbital sharing results in bond properties which differ fundamentally from those encountered in ionic (electrovalent) bonding: with the formation of the covalent bond between the hydrogen atoms (H_2 molecule formation) the bond forming capabilities of the two hydrogen atoms are saturated; the final product is a distinct molecule (H_2) rather than a giant-sized solid body which is obtained as the final product with ionic bond formation.

Covalent solid bodies, however, also do exist: they are formed if the elements involved have the capability of forming more than one bond. For example, carbon will form four covalent bonds in tetrahedral configuration - the result is diamond, a covalently bonded, three dimensional network.

DIATOMIC MOLECULES INVOLVING DISSIMILAR ATOMS

Consider the compound hydrogen chloride (in the gaseous, liquid or solid state). This compound is not ionic (because the energy state on complete ionization would be higher. Instead, bonding between hydrogen and chlorine atoms is accomplished by the sharing of electrons in a molecular orbital, thus forming a single covalent bond.

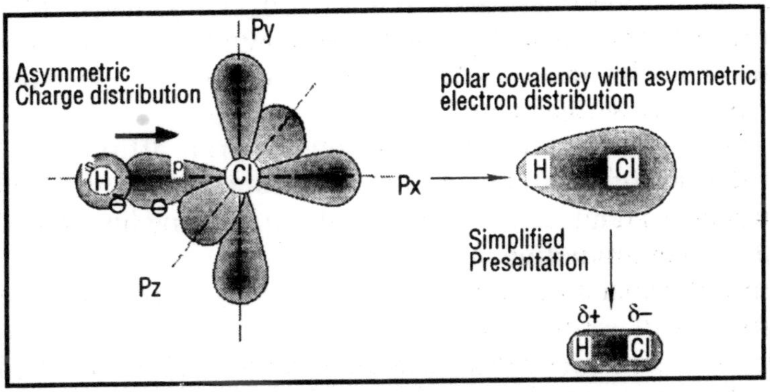

Fig. Diatomic Molecules.

The electrons involved are the 1s electron of the hydrogen atom and the unpaired 3pz electron of the chlorine atom. After spin-pairing, a typical bond results in which each atom attains a noble gas structure: in HCl the hydrogen atom is "involved" with two electrons, as in helium, and the chlorine atom with 18 electrons, as in argon. The other (inner) electrons of chlorine do not participate in the bonding; they are termed non-bonding.

According to the above considerations, the bonding of HCl may be taken as very similar to that of H_2. However, in the hydrogen molecule the electrons participating in the σ bond formation do not favour the proximity of either of the hydrogen nuclei - instead, they are equally shared. On the other hand, a preference for one nucleus (chlorine) is shown by the electrons of the σ bond in hydrogen chloride.

This is because electrons, for energetic reasons, favour the environment of the more electronegative atom; in hydrogen chloride this is the chlorine atom and, consequently, the d?bond electrons spend more time in the vicinity of the chlorine atom.

This situation results in the chlorine end of the molecule being fractionally negatively charged (δ^-) and the hydrogen end being fractionally positively charged (δ^+). We denote such an internal charge redistribution by the symbolism $H^{\delta}+ - Cl^{\delta-}$, where the δ sign indicates a partial electronic charge; the bond is said to be polar. (The hydrogen molecule does not a priori exhibit such an "asymmetric charge distribution"; it may, however, acquire a temporary polarization.)

Molecules with asymmetric electronic charge distribution have a permanent dipole moment, the value of which is given by the product of the fractional charge (δ + must be equal to δ –) and the distance of charge separation (L). Although dipole moments can be measured, their values do not allow us to directly calculate the polarity of bonds between atoms since the detailed internal charge distribution is unknown. The direction of electron drift and, to some extent, the magnitude of its effect can be estimated from the magnitude of the difference in electron affinity between the reaction partners.

Because of a limited data base on electron affinity, L. Pauling introduced a related term, the relative electronegativity. From the above considerations it should be clear that no sharp dividing line exists between ionic and covalent bonding.

We might consider a completely ionic bond to result in cases where the electron drift is such that one atom (the cation) becomes entirely deficient in one or more electrons, and the other atom (the anion) becomes correspondingly electron-rich, the bonding electrons being entirely under the influence of the latter. Hydrogen halides are σbonded and have dipole moments; the bonds are referred to as polar covalency.

The homonuclear diatomic molecules formed by the halogens, viz. F_2, Cl_2, Br_2 and I_2, are all σ-bonded systems involving spin-pairing of the various pz electrons (2pz, 3pz, 4pz and 5pz respectively). They have no permanent dipole moments.

ENERGETICS OF COVALENT BONDING

Pauling extensively treated the energetics of polar covalencies, encountered in all heteronuclear systems (such as H–Cl) which have permanent dipole moments ($H^{\delta+} - Cl^{\delta-}$). His approach visualizes the bonding to consist of two components - a pure covalency and an ionic bonding component with attraction resulting from the interaction of the fractional charges on the nuclei involved.

Pauling determines the basic covalent bonding component, for example in the formation of HCl, from the experimentally obtained bond energies associated with the molecular species of the components, H_2 and Cl_2. For this purpose he made the basic assumption that the covalent bond component between the dissimilar atoms is given by the geometric mean of the pure covalent bond energies associated with these molecular species. Thus:

Pure Covalent Bond Energy of HCI:

$$BE_{HCI} = \sqrt{BE_{H_2} \times BE_{CI_2}}$$

$$H_2 : H-H(\text{Bond Energy} = BE_{H_2} = 430\text{kJ/mole})$$

$$CI_2 : CI-CI(\text{Bond Energy} = BE_{CI} = 238\text{kJ/mole})$$

$$BE_{HCI} = \sqrt{430 \times 238} = 320\text{kJ/mole}$$

The ionic contribution (Δ) to the polar covalent bonding is then obtained from the difference between the experimentally determined bond energy (for HCl, for example), which obviously must contain both components, and the calculated pure covalent bond energy:

$$\Delta = [BE_{HCI}(\text{experimental})] - BE_{HCI}[(\text{Theoeretical}$$

$$-\text{covalent})]$$

$$(\text{The experimentally determined } BE_{HCI} = 426\text{kJ/mole.})$$

$$\Delta = 426 - 320 = 106\text{kj}$$

In connection with the presently discussed work, Linus Pauling established the now generally used electronegativity scale, or, better, the scale of relative electronegativities. This scale, included in the Periodic Table of the Elements, is extremely helpful since values of the electron affinity are known thus far for only very few elements.

The electronegativity (x) scale lists the relative tendency of the neutral elements to attract an additional electron. (The values of x are conventionally listed in electron Volts.) The scale listed was obtained by arbitrarily fixing the value of x_H = 2.2. Pauling obtained the values for the other elements by relating differences in electronegativity of reaction partners to the fractional ionic character (ionic bonding component) of the bond established between them:

$$\text{Experimental } BE_{AB} = \sqrt{BE_{AA} \times BE_{BB}} + k(X_A - X_B)^2$$

or

$$\Delta = 96.3(X_A - X_B)^2 \text{KJ}$$

According to Pauling, the bonding character between two different elements may be defined as:

Ionic bonding for: $\Delta X > 1.7$;Covalent bonding for:$\Delta X < 1.7$

The fractional ionicity of polar covalent bonding as listed in the P/T is obtained by the relationship:

$$\%\text{ionic bonding} = (1 - e^{-0.25(X_A X_5)^2}) \times 100$$

BONDING IN POLYATOMIC MOLECULES

In the formation of covalent bonds between atoms in polyatomic molecules, the conditions that atomic orbitals distort so that maximum overlap may be achieved when bonding occurs produces more extensive changes of orbital geometry than is the case with diatomic molecules.

It will be remembered that the changes which result in a d bond formation in diatomic molecules are a distortion of the s orbitals or p lobes so that strongest bond formation results from the maximum overlap between the two joined nuclei.

For polyatomic molecules extensive alterations in the spatial disposition of atomic orbitals do occur, and very often the unique orbital geometry of the original atomic orbitals is completely lost. It is sometimes convenient to view these alterations as occurring in each atom prior to bonding by a process involving the mixing, or hybridization, of atomic orbitals.

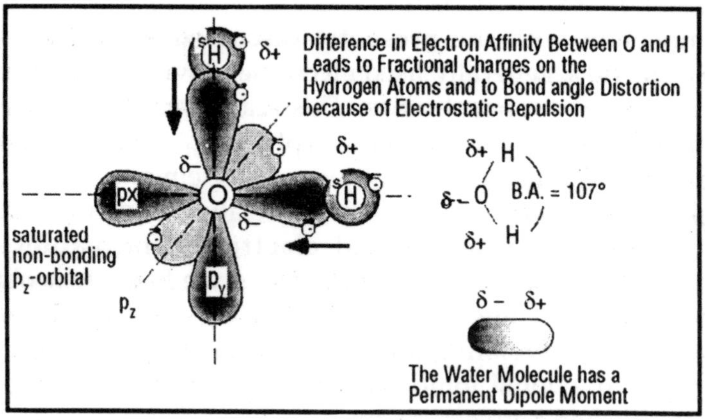

Fig. Bonding In Polyatomic Molecules.

BONDING INVOLVING CARBON

In methane (CH_4) the four outer, or valence, electrons of carbon are shared with the electrons of hydrogen; there is spin-pairing (resulting in bond formation) between each individual hydrogen electron and one of the carbon valence electrons.

The noble gas structure is thus attained by each nucleus: the carbon nucleus "sees" eight outer electrons and each H nucleus "sees" two electrons. Consider now the orbitals which are involved in more detail. Each hydrogen has one spherical valence orbital (the 1s orbital) containing one electron, and covalent bond formation results from its distortion, overlap with the valence orbitals of carbon and spin-pairing.

According to the Aufbau principle and Hund's rule, in its ground state carbon has two electrons in the filled K shell and four electrons in the L shell: two 2s electrons and two 2p electrons in singly occupied orbitals which are capable of covalent bond formation.

This configuration provides, in principle, only two orbitals (2px, 2py) for covalent bond formation. However, with two covalencies carbon will not yield the desirable octet configuration.

Such a configuration can be obtained if one of the two 2s electrons is "promoted" into the empty 2pz orbital since this process results in four singly occupied orbitals - all of which, being singly occupied, are capable of bond formation.

The dissimilar singly occupied orbitals can assume (and therefore will assume) upon bond formation a lower energy configuration involving a process called hybridization whereby the four orbitals (of two different types) hybridize into four identical orbitals of maximum equal spacing from each other.

Thus the hybridized orbitals (sp^3 hybrids) are lobes emanating from the carbon atom into the corners of a tetrahedron, forming bond angles of 109°28′. sp^3 hybridization is characteristic for carbon; however, other types of hybridization, sp^2 and sp, are encountered in other elements as well.

Boron, for example, will tend to promote one of its two 2s electrons into a 2p state and, by hybridization, form three equivalent sp^3 orbitals which assume planar orientation with band angles of 120°. Beryllium forms sp hybrid orbitals of linear orientation (the bond angle is 180°). All hybrid orbitals are capable of σ bond formation.

The great diversity of carbon compounds (it forms more compounds than all the rest of the elements in the Periodic Table) can be attributed to the hybridization capability of the carbon orbitals (sp^3, sp^2 and sp). As a consequence, carbon forms not only axisymmetric d?bonds (through axial orbital overlap), but also σbonds (through lateral orbital overlap).

In the compound ethane ($H_2C = CH_2$), the interactive carbon atoms undergo sp^2 hybridization for a dbond by overlap of two sp^2 hybrid orbitals and form, in addition, a σ bond by lateral overlap of the remaining non-hybridized π orbitals. (Double bonds involve one σ and one π bond.) In acetylene ($HC \equiv CH$), we find sp hybridization and, by axial overlap, bond formation as well as lateral overlap of the remaining non-hybridized px and py orbitals which form two bonds.

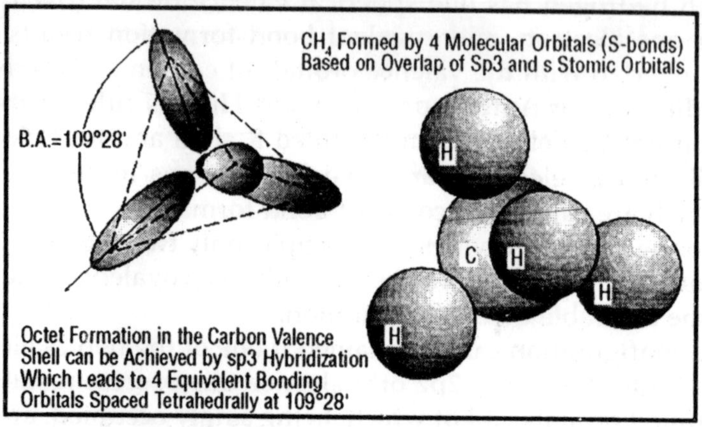

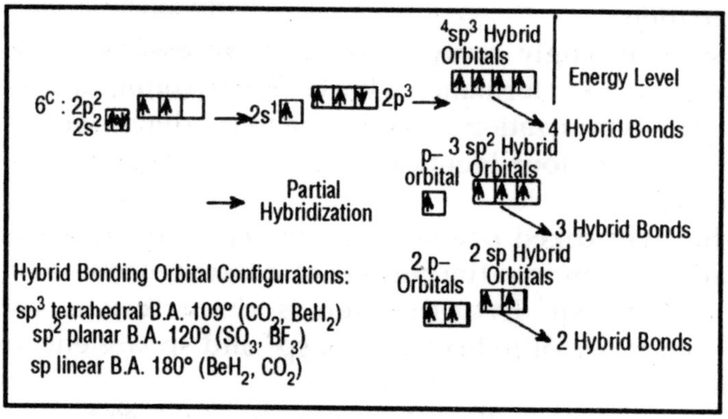

Fig. Hybrid Bonding Orbital Configuration.

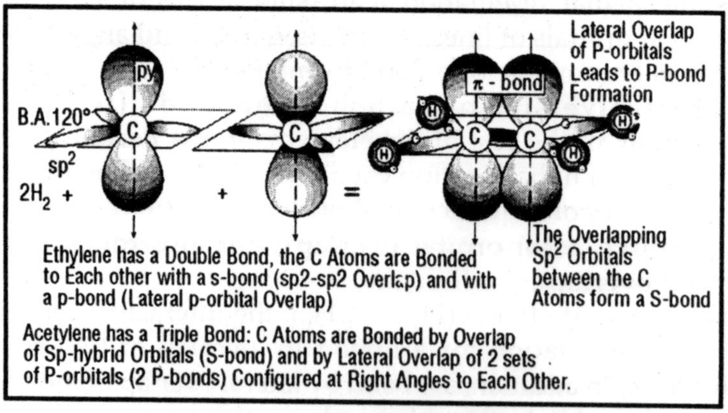

Fig. Triple Bond of Acetylene.

METALLIC BONDING

The stability of both covalent and metallic bonds may be regarded as arising from the potential energy lowering experienced by valence electrons under the influence of more than one nucleus. In metals, where the valence electrons are not as tightly bound to their ion cores, we cannot expect the formation of strong electron-pair bonds.

The bond energies of known diatomic molecules of metallic elements are, in fact, smaller than those consisting of nonmetallic elements: 104 kJ/mole for Hg_2. Only diatomic molecules of the semi–metals have relatively high binding energies (385 kJ/mole for As_2, 293 kJ/mole for Sb_2 and 163 kJ/mole for Bi_2). These values reflect multiple bonding. Much greater stability is possible in larger aggregates of atoms such as bulk metals.

The known properties of metals, such as low electrical resistance and malleability, support the conceptual view that the valence electrons in metals never remain near any particular atom very long, but drift in a random manner through the lattice of ion cores. We may therefore visualize metals as a lattice of ion cores being held together by a gas of free electrons

SECONDARY (VAN DER WAALS) BONDING

Primary bonding (ionic, covalent and metallic) is strong and the energies involved range from about 100 to 1000 kJ/mole. In contrast, secondary bonding is weak, involving energies ranging from about 0.1 to 10 kJ/mole. While this type of bonding, also referred to as "residual", is weak, it is essential in the functioning of our environment. Coke would likely be gaseous and not a bubbly, refreshing brew were it not for secondary bonding, nor would catalytic converters function.

The energy difference between the liquid and vapour states of a given system is given by the heat of vaporization, i.e. the heat required to convert a given liquid into a vapour (normally) at the boiling point temperature at 1 atm pressure.

The energy difference is due to intermolecular attraction between molecules at close distance of separation. This phenomenon of attraction through secondary bonding can best be considered between a single pair of molecules, but recognizing that the forces are of longer range. Four types of intermolecular forces can be identified.

- *Dipole-Dipole Interaction*: Molecules with permanent dipoles (such as water, alcohol and other organic compounds with functional groups) exert a net attractive force on each other as a result of varying degrees of alignment of oppositely charged portions of the molecules. For two polar molecules with a dipole moment of

(μ) separated by a distance of (r), the energy of attraction can be quantified as:

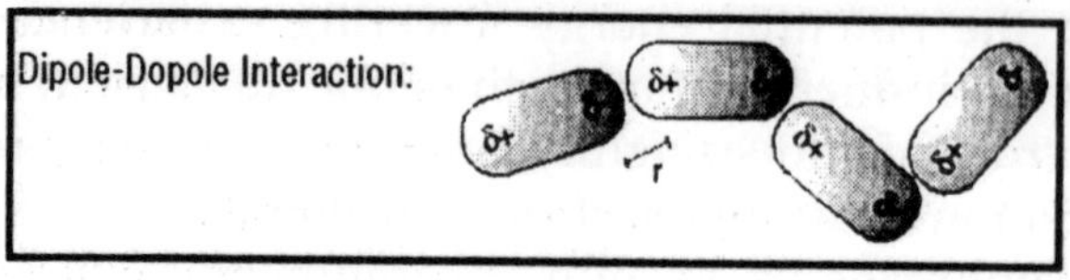

Fig. Dipole - Dipole Interaction.

$$E_{Dipole-Dipole} = \frac{2}{3}\frac{\mu^4}{Dr^6} \times \frac{1}{kT}$$

where μ is the dipole moment, r is the distance of approach of the oppositely charged molecular portions, k is the Boltzmann constant and T is the absolute temperature in K.

The molar energies of attraction associated with dipole-dipole interaction range from 0 to about 10 kJ/mole. These forces are primarily responsible for the liquid state (at room temperature) of most polar organic molecules; they are a contributing factor for H_2O to be liquid at room temperature, and are responsible for alcohol being a liquid.

- *Dipole-Induced Dipole Interaction*: A dipole in one molecule can interact with and polarize the electrons of a neighboring non-polar molecule, thus generating an induced dipole which will experience an attractive force with the polarizing, polar dipole. P. Debye showed that in a molecule with a "polarizability" of (α) the attractive potential arising from dipole-induced dipole interaction is given as:

$$E_{Dipole-Induceddipole} = -\frac{2\alpha\mu^2}{Dr^6}$$

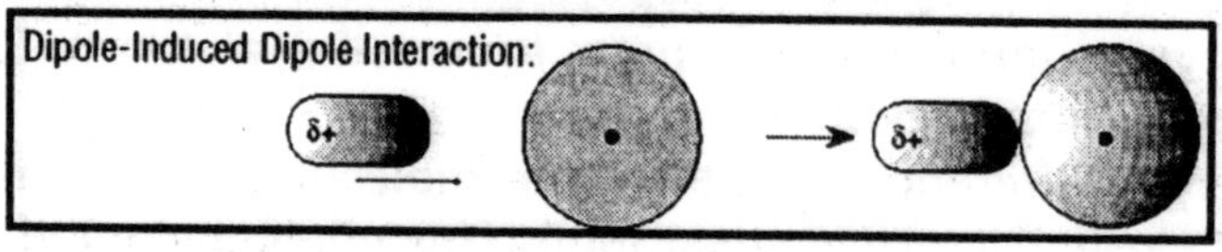

Gig. Dipole-Induced Dipole Interaction.

Induced dipole interaction is important in aqueous solutions and very effective during adsorption of inert molecules on active solid substrates.

LONDON DISPERSION FORCES

It is a well-known fact that all substances, including rare gases and hydrogen, assume liquid state at finite temperature, an indication of the existence of attractive interatomic and intermolecular forces, even in the absence of permanent dipole systems. The origin of this force has been proposed by F. London in 1930. Accordingly, orbiting electrons will at any instance generate a "temporary dipole", the configuration of which changes as the electrons move.

Since all atoms of a given system similarly experience temporary instantaneous dipole moments, their effect is expected to be cancelled because of the statistically random orientation of dipoles. It is evident that, should the dipoles be synchronized in a given assembly of atoms, then a net attractive force would result.

But since such an attractive force constitutes a lowering of the energy of a given system, synchronization can and will take place because all systems will attempt to assume minimum energy configuration. The London dispersion force can be formulated as:

$$E_{London} = -K\frac{\alpha^2}{r^6}$$

The attractive London forces are small, as manifested by the very low boiling points of the smaller rare gases, of hydrogen and nitrogen.

Hydrogen Bonding

The short and long range dipole interactions calculated from molecular dipole moments are inadequate in explaining a multitude of phenomena in organic as well as some inorganic systems. L. Pauling studied such London Dispersion (Fluctuating Dipole Interaction) Electron motion in atomsgenerates fluctuating dipoles. Since the polarization is random attractions are cancelled by repulsions.

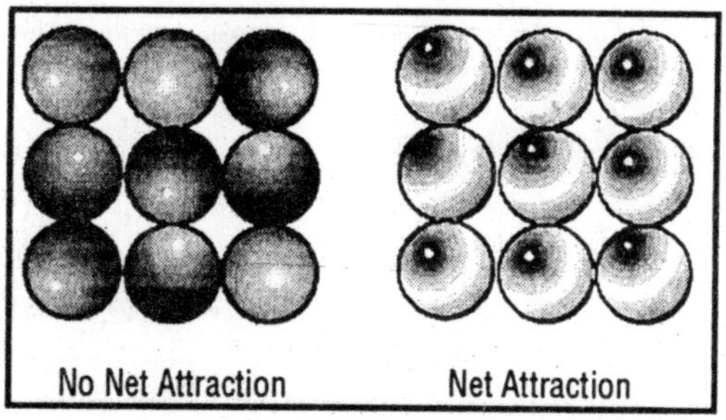

Fig. Fluctuating Dipole Interaction.

London

Upon synchronization of electron motion inter atomic attraction will be established in an assembly of atoms. Since this attraction constitutes a lowering of the energy of the system - attraction by synchronization will take place - as all systems attempt to assume lowest energy configuration phenomena and concluded the existence of highly specific attractive interaction between hydrogen that is acidic (carries a fractional positive charge) and the elements O, F, N and, to a lesser extent, S in both organic and inorganic molecules.

This interaction, which can as yet not be formulated, is referred to as hydrogen bonding; its magnitude, ranging up to 40 kJ/mole, is significantly larger than that of any other secondary bonding type. Hydrogen bonding is considered instrumental in controlling most properties of water, is a key element in the structure of nucleic acid and thought to be an essential component in memory functions of the human brain.

Chapter 8

Bonding and Band Structure

Characteristic properties of metals include:

- Electrical conductivity.
- Opaqueness and.
- Malleability.

A very simple model in which the metallic crystal is viewed as a lattice of positive ions surrounded by a "gas" of free electrons provides a crude understanding of the first and third properties. If the crystal has a gas of free electrons, it is easy to see why the application of an electric field will result in the motion of these electrons and thus for the high electrical conductivity.

This model also allows one to explain the malleability of metals. When a metallic crystal is subject to forces that displace one plane of atoms with respect to another, the environment of the charged species is left unchanged. In contrast, the displacement of neighboring planes in an ionic crystal as a result of a distorting force will lead to cleavage, largely because of changes in the interaction of the charged species.

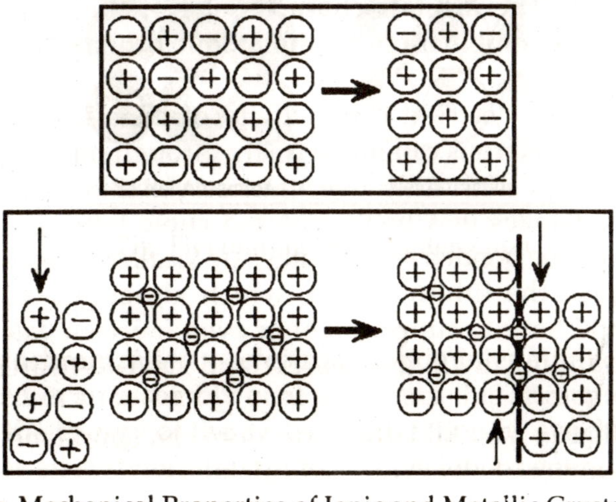

Fig. Mechanical Properties of Ionic and Metallic Crystals.

The free electron gas theory (Drude–Lorentz) also explains in principle the nature of the attractive bonding forces which hold the metallic ions (the metal) together: the crystal is held together by electrostatic forces of attraction between the positively charged metal ions and the non-localized, negatively charged electrons - the electron gas. The theory in its original form assumes that the classical kinetic theory of gases is applicable to the electron gas; mutual repulsion between electrons was ignored and the electrons were expected to have velocities which are temperature dependent according to a Maxwell-Boltzmann distribution law.

While the Drude-Lorentz theory of metallic bonding was considered a useful model, several shortcomings soon became apparent. The most notable failure consisted of the unexplainable discrepancy between the observed and predicted specific heats of metals (energy in the form of heat, required to increase the temperature of 1 g of a given metal by 1·C).

The D-L theory predicted much larger specific heats than are observed (because the Maxwell-Boltzmann energy distribution has no restrictions as to the number of species allowed to have exactly the same energy). [If there are restrictions as to the number of electrons with identical energy (Pauli exclusion principle!), we have to apply a different form of statistics (Fermi-Dirac statistics).]

THE ENERGY BAND THEORY BASED ON FORMATION OF MOLECULAR ORBITALS

A different model, and one that is more closely related to the models of the chemical bond discussed earlier, is the band model. This model, proposed by Bloch before the development of the molecular-orbital approach to chemical bonding, is actually a molecular-orbital model of metallic crystals. The orbitals characteristic of the whole crystal are obtained as linear combinations of the atomic orbitals of the individual atoms.

From the large electrical conductivity of metals, it appears that at least some of the electrons can move freely through the bulk of the metal. Since even lithium with only one valence electron has eight nearest neighbors (crystallizes in a body-centered cubic lattice), it is clear that the atoms can not be bonded to each other by localized electron-pair bonds (for in that case the lithium atom, which altogether has only three electrons, would have to supply eight or even fourteen valence electrons to establish bonding with the nearest and next-to-nearest neighbour atoms).

The fact that in a metal lattice one atom can simultaneously interact with a large number of others can be explained by considering that metals constitute an extreme example of delocalized bonding. Conditions in metals are not as easily assessed as in organic hydrocarbon chains because metal lattices are three-dimensional structures. However, several important results from the theory of metallic bonding can be understood on a highly simplified model in which the three-dimensional lattice is replaced by a one-dimensional system.

Consider the formation of a linear array of lithium atoms from individual lithium atoms:

$$\text{Li} \rightarrow \text{Li–Li} \rightarrow \text{Li–Li–Li} \rightarrow \text{Li–Li–Li–Li} \rightarrow ...$$

The first stage is the formation of a lithium molecule, Li_2, which is comparable to the hydrogen molecule, H_2. The two lithium atoms are bound together by a pair of valence electrons; each lithium atom supplies its 2s-electron which, through orbital overlap, forms a covalent molecular bond. The molecule formed occurs in lithium vapour

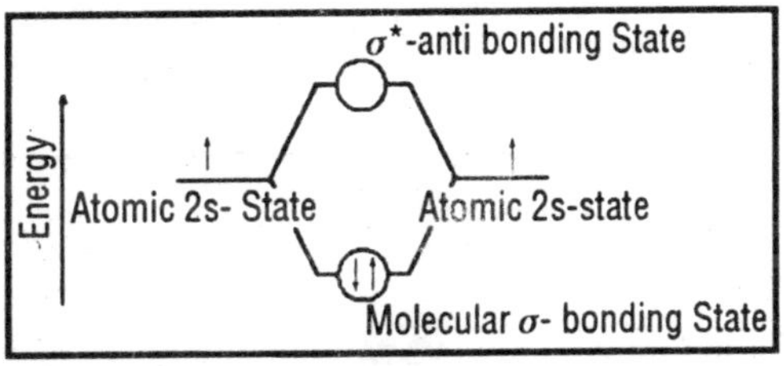

Fig. Formation of Li2 Molecule.

and differs from the H_2 molecule only by its greater interatomic distance (Li–Li = 2.67 × 10–10m, H–H = 0.75 × 10^{-10} m) and smaller energy of formation. The larger separation and the consequent reduction in bond energy can be attributed to the size of the Li^+ atomic core.

In the determination of molecular orbital states by linear combination of atomic orbitals (LCAO) we obtain more than one solution from the Schrödinger equation. We have so far concentrated mostly on the low energy states, leading to dor σor π bonds.

We have largely ignored the so-called antibonding states which in covalent molecules have energies higher than the original energy states of the isolated reaction partners (presently Li) and, therefore, remain unoccupied (or empty) in two-atomic molecules. In any multi-atom system (such as a piece of metal), the total number of molecular orbital states (electronic states) is always the same as the total number of original atomic states (conservation of electronic states). However, not all of these states are necessarily occupied by electrons.

Consider the hypothetical linear molecule Li_3. Since the valence electron cloud is spherical, the central lithium atom can not give preference to either of its neighbors. The three atomic valence electron clouds overlap to form one continuous distribution and two others with nodes (antibonding states), i.e., three molecular orbitals in all.

As the length of the chain is increased, the number of electronic states into which the atomic 2s state splits also increases, the number of states always equaling the number of atoms. The same occurs when lithium chains are placed side-by-side or stacked on top of each other, so that finally the space lattice of the lithium crystal is obtained.

It is of great significance that these electronic states have energies which are bounded by an upper and lower limiting value. Within these limits the states form an energy band of closely spaced values (one gram of lithium contains nearly 10^{23} atoms). Similarly, energy bands can also result from overlapping p and d orbitals.

The electronic states (orbitals) within an energy band are filled progressively by pairs of electrons in the same way that the orbitals of an atom were filled in accordance with the Pauli principle. This means that for lithium the electronic states of the 2s band will be exactly half-filled.

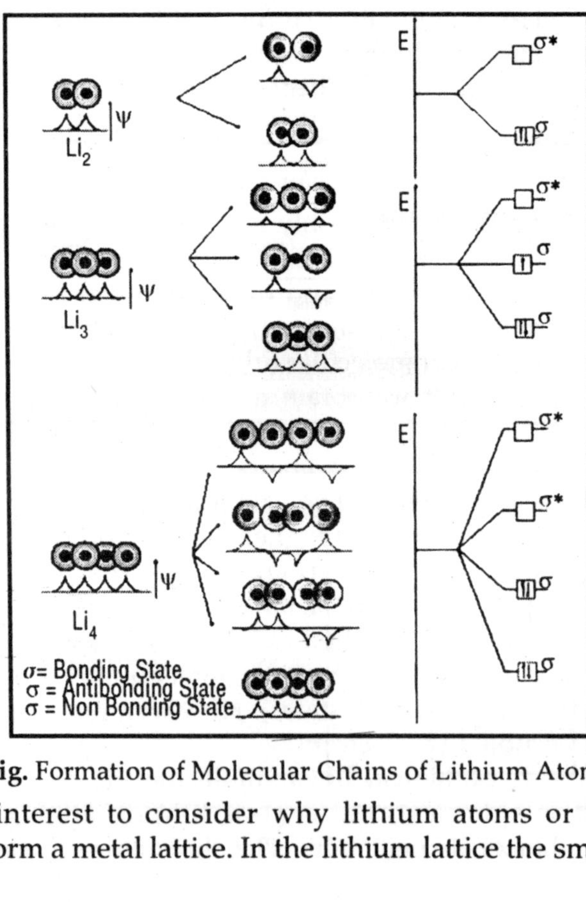

Fig. Formation of Molecular Chains of Lithium Atoms.

It is of interest to consider why lithium atoms or Li_2 molecules combine to form a metal lattice. In the lithium lattice the smallest distance between neighboring atoms is 3.03×10^{-10} m, which is larger than in the Li_2 molecule. This reflects the fact that bonds between pairs of atoms in the metal are weaker than they are in the molecule.

Nevertheless, the metallic form of lithium is more stable than the molecular form because in the metal one atom has many more neighbors than in the Li_2 molecule. As a result, the binding energy per gram atom of lithium (i.e., per 6.92 g of lithium) is 163 kJ for the metal lattice, but only 56 kJ for one mole of molecule. [The possibility of hybridization (first advanced by L. Pauling to explain metallic bonding) is also a likely factor for the formation of metallic bonds. Thus strong bonds can be formed when the valence electron clouds become concentrated along the direction in which the bonding partners are situated. According to Pauling the situation

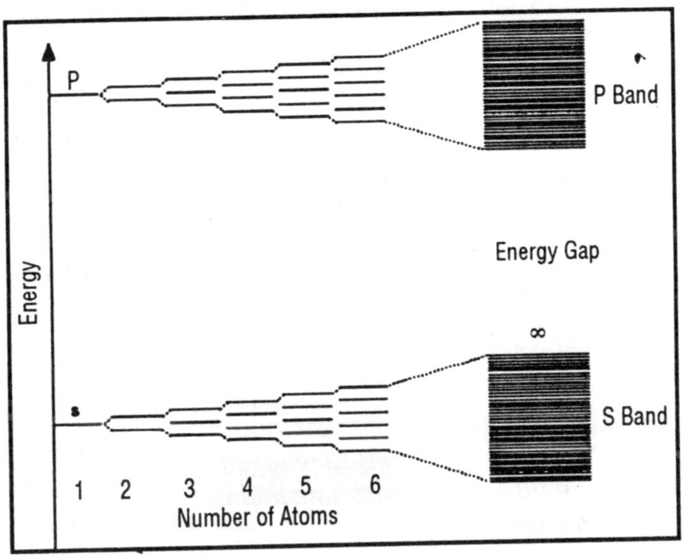

Fig. Formation of Energy Bands from Energy Levels of Constituent Atoms

can be described by mesomeric limiting formulae, just as in the theory of unsaturated carbon compounds. One reason given for hydrogen not forming a metal is its inability to hybridize.]

The appearance of bands of allowed electron energies relates to the overlap of electron Ψ functions. The width of each energy band is a function of the crystal structure because it determines the number of nearest neighbors in the crystal.

Different metal atoms with the same valence, particularly elements in the same group of the periodic table, can often replace each other in arbitrary proportions without altering either the lattice type or the structure of the energy bands. This explains why such metals tend to form a complete series of solid solutions. Metallic alloys consist of such solid solutions or of heterogeneous mixtures of such solutions. Within certain limits, even metal atoms of different valence can be interchanged in a lattice.

BAND STRUCTURE OF METALS

According to previous reasoning, the 2s band has N states (N = number of atoms) and accommodates n 2s electrons (where n is the number of electrons per atom in the 2s state times N). Thus, this band has only half of the states filled since each state can accommodate two electrons of opposite spin (Pauli exclusion principle).

In accordance with the Aufbau principle, the lowest energy states of the band are filled first and the upper states remain empty – but can readily be occupied by electrons upon thermal excitation or the application of an electric field.

Since the width of the energy band is of the order of a few volts, spacings of states within the band are of the order of ~10^{-20} eV (1 eV = 1.6 × 10^{-19} J), electrons can readily acquire the energy necessary to move into excited states, be accelerated, and move through the metal as conducting electrons. Partly filled bands thus constitute conduction bands.

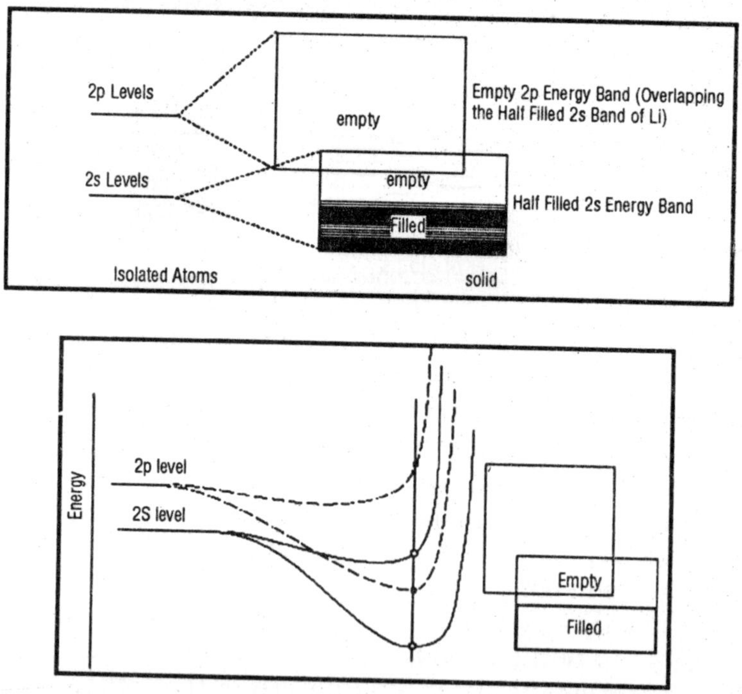

Fig. Schematic Energy Band Configuration for Li.

The conduction mechanism in Mg, for example, appears complicated by the fact that each 3s state of the valence shell in the atoms is doubly occupied ($3s^2$). Thus the 3s band must be filled completely and no electronic conduction would in principle be expected.

Electronic conduction, however, is observed because of a partial overlap of the 3s and the empty 3p bands. With this overlap, electrons can be activated into empty 3p states and exhibit conduction, as in the partly filled s band in Li.

BAND STRUCTURE OF INSULATORS AND SEMICONDUCTORS

The same band model can also account for the lack of conductivity in covalent crystals, such as diamond. In the diamond crystal, with a two-electron bond between every atom and each of its neighbors, the bonding molecular orbitals form a band at much lower energies than the energy of the atomic sp^3 hybrid orbitals, and the antibonding orbitals form a band at energies much higher than the energy of the atomic orbitals.

Also, since each atom contributes four orbitals and four electrons, there are just enough electrons to fill the bonding orbitals (corresponding to an electron pair bond joining every atom to each of its neighbors) and the only vacant orbitals are those in the high energy band of antibonding orbitals.

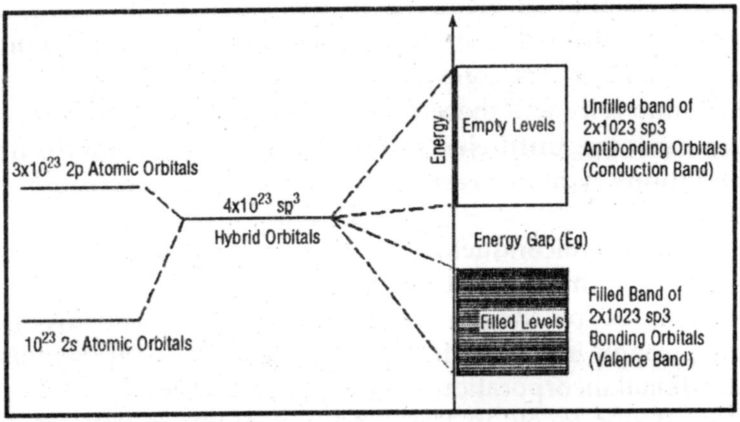

Fig. Band Structure of Insulators and Semiconductors (Molecular Crystals); the Conditions Depicted Reflect a Molar Crystal of Carbon (diamond).

Both insulators and semiconductors have the same basic band structure – the primary difference is the width of the forbidden energy gap (E_g) between the valence and the conduction band.

Insulators, generically, are materials with very high resistivity, comprising glasses, polymers, refractories, composites, liquids and gases. In the present context, an insulator is a molecular crystal, such as diamond (C) or sapphire, with a band gap (E_g) in excess of 4 eV (arbitrary value).

Generally such materials will not conduct electricity since their valence band is filled and the energy required to transfer electrons from the valence band to the empty conduction band is far in excess of both the thermal energies at room temperature and the energy provided by radiation of the visible spectrum (~2 eV).

Therefore, insulators (in single crystal form) are normally transparent (colorless); however, if light is excessively or totally scattered at internal heterogeneities (such as grain boundaries), they may be translucent and even opaque. It should also be recognized that impurities (Cr^{3+} in Al_2O_3) or particular point defects (colour centers) may impart a colour to the transparent insulator crystals. The colour arises because of partial absorption of white light and selective transmission of the other portions of the visible spectrum.

SEMICONDUCTORS

The conventional semiconductors, silicon (Si) and germanium (Ge), have a band gap (Eg) of 1.1 and 0.7 eV respectively and therefore absorb visible radiation; they are opaque. Considering the statistical nature of the thermal energy distribution in the solid matrix (Maxwell-Boltzmann), a significant number of electrons in the valence band will, at room temperature, acquire sufficient energy to cross the existing energy gap and thus provide for semiconductivity.

The conductivity will therefore increase with temperature, contrary to metallic systems, until electron scattering effects, due to increased lattice vibrations (which decrease the mobility of electrons), begin to dominate.

The value of semiconductors for solid state device fabrication lies in the fact that the number and type of conducting electric charge carriers [electrons are n-type (negative), holes are p-type (positive)] can be controlled through incorporation of appropriate dopant elements. Thus the substitutional incorporation of Group V elements (Sb, As, P) provides for shallow donor levels in the band gap at about 0.01 eV from the conduction band.

The substitutional incorporation of Group III elements (B, Al) generates acceptor levels in the band gap at about 0.01 eV from the valence band. The two types of impurities are almost completely ionized at room temperature and give rise to extrinsic n-type and p-type conductivity – the basis for the formation of junction devices such as diodes and transistors.

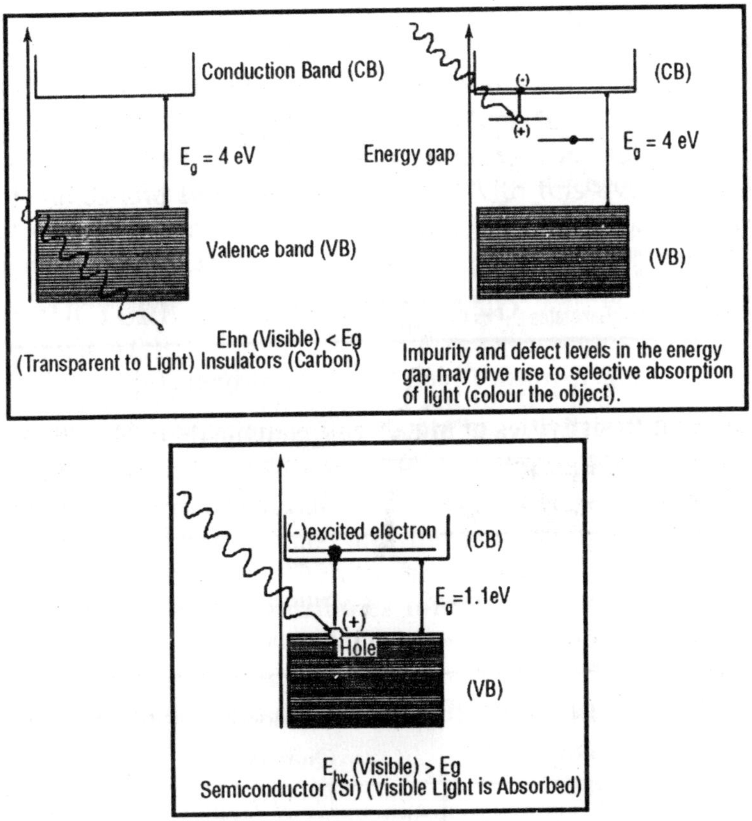

Fig. Optical Bebavior of Insulators and Semiconductors.

Of increasing importance are compound III–V (adamantine) semiconductors, such as GaAs, InSb, InP and GaP (compounds of Group III and Group V elements). Together these compounds provide eight valence electrons and, by sp^3 hybridization, are able to form a diamond-like, covalent crystal structure with semiconductor properties. These compounds (GaAs, for example) exhibit electron mobilities which are higher than those of silicon and, therefore, are of considerable interest for advanced device technology.

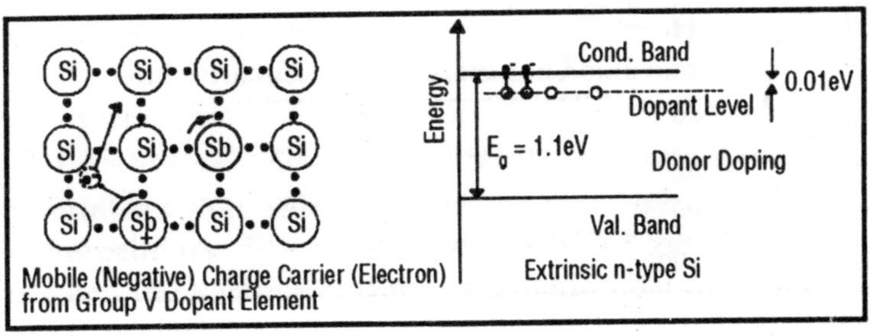

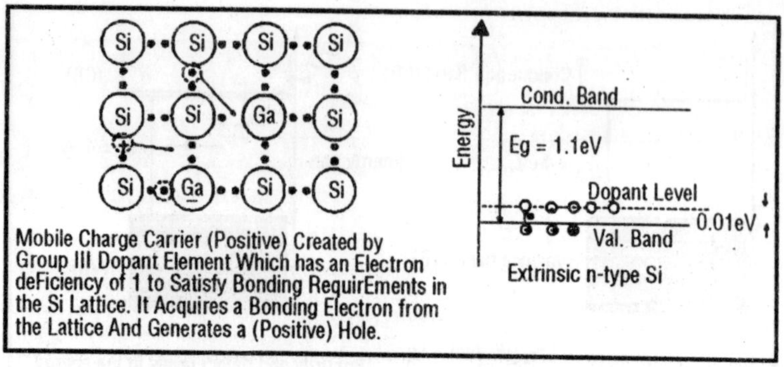

Fig. Extrinsic P- and N-type Semiconductors.

Table. Electrical Resistivities of Metals and Nonmetals at 20°C ELECTRICAL

Metals	Resistivity, 10–8 ohm–m	Nonmetals	Resistivity, ohm–m
Silver	1.6	Semiconductors	
Copper	1.67	Silicon	1000.0
Gold	2.3	Germanium	0.09
Aluminum	2.69	Insulators	
Magnesium	4.4	Diamond	10^{10}–10^{11}
Sodium	4.61	Quartz	1.2 x 10^{12}
Tungsten	5.5	Ebonite	2 x 10^{13}
Zinc	5.92	Sulfur	4 x 10^{13}
Cobalt	6.24	Mica	9 x 10^{13}
Nickel	6.84	Selenium	2 x 10^{14}
Cadmium	7.4	Paraffin wax	3 x 10^{16}
Iron	9.71		
Tin	12.8		
Lead	20.6		
Uranium	29		
Zirconium	41		
Manganin	44		
Titanium	55		
Lanthanum	59		
96%Iron–4%Si	62		
Cerium	78		
Nichrome	100		

Chapter 9

Theory of Reaction Rates

We can readily understand that chemical reactions normally are preceded by collisions of atoms, ions, or molecules. However, the rate of such collisions in solids, liquids and gases is so great that all reactions would be very rapid were it only necessary for collisions to occur. "Chemical" reactions will not proceed more rapidly than molecular collisions allow, but many reactions proceed much more slowly.

It is thus apparent that not every molecular collision leads to reaction. We have also seen from earlier examples that the driving force for physical as well as chemical reactions is the (free) energy change - which must be negative for reactions to occur spontaneously. However, while providing a "go/no-go" answer, this criterion alone cannot give us any information concerning the rate at which reactions occur, nor can it tell us which factors influence the reaction velocity.

Rates at which reactions occur vary considerably. For example, the nuclear reaction:

$$U_{92}^{238} \rightarrow Th_{90}^{234} + \alpha \text{ is 50\% completed } (\tau_{1/2}) \text{ after } 5 \times 10^9 \text{ years.}$$

$$\tau_{1/2} = 5 \times 10^9 \text{ years}$$

On the other hand, the chemical reaction:

$$SO_4^{--} + H^+ \rightarrow HSO_4^- \quad \text{is 50\% completed after about } 10^{-4} \text{ s.}$$

$$\tau_{1/2} = 3 \times 10^{-4} \text{ Seconds}$$

Reaction kinetics (rate theory) deals to a large extent with the factors which influence the reaction velocity. Take corrosion (rusting of iron), for example. We all know that it requires air and water to provoke rusting and we also know, much to our sorrow, that rusting proceeds much more rapidly near the ocean where salt is present.

We also know that the rate of rusting depends strongly on the composition of iron (pure Fe and steel corrode much less rapidly than cast iron, for example).

It is primarily kinetic studies which lead to the elucidation of chemical reactions and, in the case of corrosion, to the development of more corrosion resistant materials.

The principal experimental approach to the study of the reaction process involves the measurement of the rate at which a reaction proceeds and the determination of the dependence of this reaction rate on the concentrations of the reacting species and on the temperature.

These factors are grouped together in the term reaction kinetics and the results for a given reaction are formulated in a rate equation which is of the general form:

Rate = k(T) x function of concentration of reactants

The quantity k(T) is called the rate constant and is a function only of the temperature if the term involving the reactant concentrations correctly expresses the rate dependence on concentration. Thus the experimental information on the reaction process is summarized in the rate equation by the nature of the concentration function and by the value and temperature dependence of the rate constant.

EXPERIMENTAL METHODS IN REACTION RATE STUDIES

Since we know that chemical reactions are temperature dependent, kinetic investigations will require rigorous temperature control (thermostats).

Furthermore, we have to be able to observe and investigate concentration changes of reactions and products.

For this purpose it frequently is customary to interrupt a reaction (for example, by quenching - the abrupt lowering of temperature) and to make a chemical analysis.

In other instances, particularly for very fast reactions, a direct measurement of concentration changes is physically impossible. In such cases it is necessary to quantitatively follow reactions indirectly, through the accompanying changes in specific physical properties such as:

- Electrical conductivity
- Optical absorption
- Refractive index
- Volume
- Dielectric constant

as well as by other means. In the last twenty years, for example, the use of isotopes (as "tracer" elements) has become a valuable tool for the study of reaction kinetics (in slow reactions).

CONCENTRATION DEPENDENCE OF REACTION RATES

Normally experimental data of kinetic investigations are records of concentrations of reactants and/or products as a function of time for constant temperatures (taken at various temperatures).

Theoretical expressions for reaction rates (involving concentration changes) are differential equations of the general form:

$$\frac{dc}{dt} = f(c_1^m, c_2^n, c_3^o ...)$$

where (c) are concentration terms which have exponents that depend on details of the reaction.

If we want to compare the theory with the experiment, it is therefore necessary to either integrate the theoretical laws or to differentiate experimental concentration vs time data.

The rate laws are of importance since they provide analytical expressions for the course of individual reactions and enable us to calculate expected yields and optimum conditions for "economic" processes.

In most instances the differential rate equation is integrated before it is applied to the experimental data. Only infrequently are slopes of concentration vs time curves taken to determine dc/dt directly.

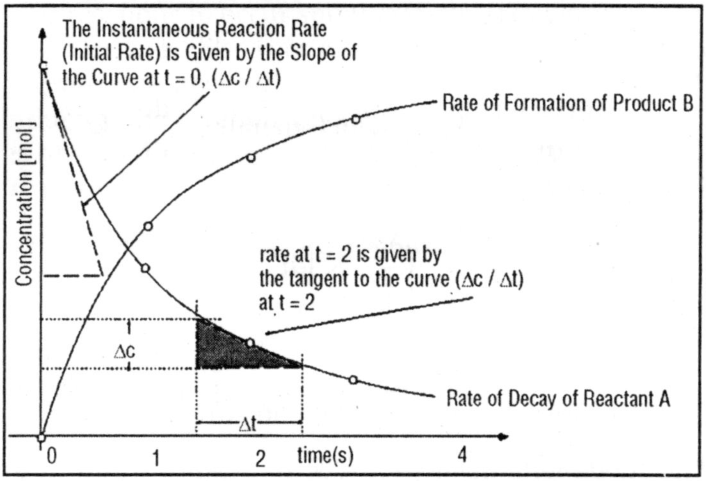

Fig. Variation with Time of the Concentration of a Reactant (A) and of a Product (B).

Let us look at a simple reaction, the decay of H_2O_2 to water and oxygen:

$$2H_2O_2 \rightarrow 2H_2O + O_2$$

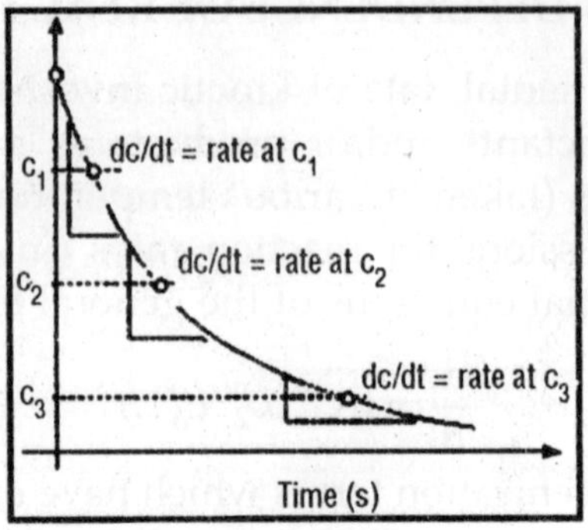

Fig. Decay Curve (dc/dt vs. c) for H_2O_2.

Experimentally we obtain a curve by plotting the concentration of remaining H_2O_2 in moles/liter, normally written $[H_2O_2]$, as a function of time.

The rate is then given by the time differential:

$$\text{Rate}\left(\frac{\text{moles}}{\text{liter}\cdot\text{s}}\right)=\frac{-d[H_2O_2]}{dt}$$

If we now determined and plotted the variation of the reaction rate, $[-d[H_2O_2]/(d\ \text{time})]$, with concentration, $[H_2O_2]$, we would find a straight line.

Thus:

$$-\frac{d[H_2O_2]}{dt}=k[H_2O_2]\text{or Generally},-\frac{dc}{dt}=kc$$

Upon integration:

$$\int\frac{dc}{c}=K\int dt$$

we have:

ln c = – kt + const.

Taking c = co for t=0, we get (const = ln co) and:

ln c = – kt + ln co

or:

$$c=c_o e^{kt}\ \text{and}\ k=\frac{1}{t}\times \text{In}\frac{c_o}{c}$$

$$k=\frac{2.3}{t}\times\log\frac{c_o}{c}\ \text{and}\ k=\frac{2.3}{t_2-t_1}\times\log\frac{c_1}{c_2}$$

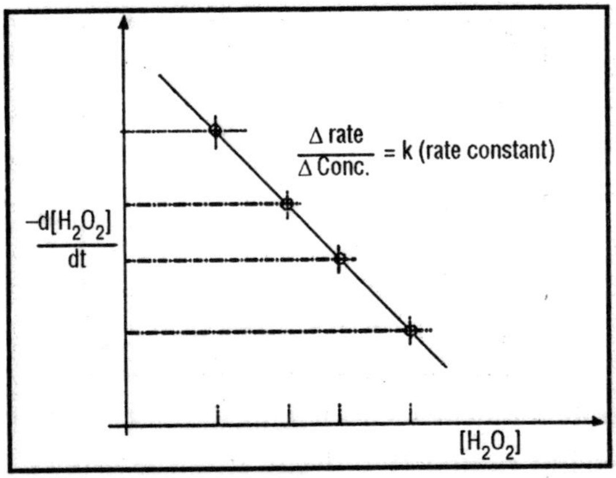

Fig. Concentration Dependance of the Rate of Reaction.

From the above we see that for decay of H_2O_2 a plot of (ln c) vs. t results in a straight line. This behaviour is characteristic of first order reactions in which the concentration exponent (n) is "one":

$$-\frac{dc}{dt} = kc$$

First Order Reactions: Radioactive Decay

From the Relationship

$$k = \frac{2.3}{t_2 - t_1} \times \log \frac{c_1}{c_2}$$

which relates the rate constant (k) to the concentration change (Δ c) for the time interval (Δ t), we can show that (for first order reactions) the time required to complete a reaction to 1/2 or 1/4 or any fraction of the initial concentration is independent of the initial concentration of reactant present.

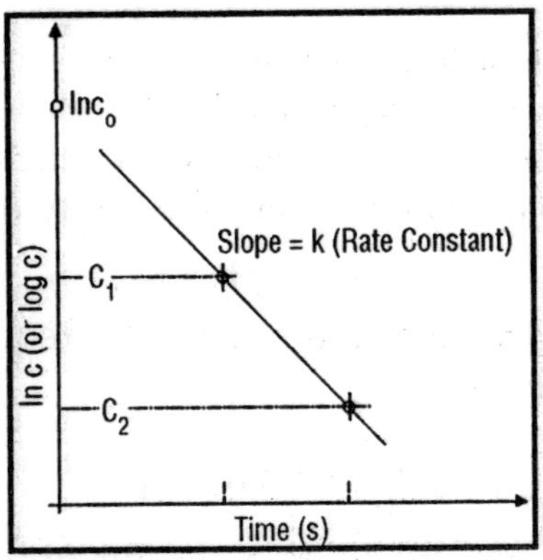

Fig. Plot of ln c vs. Time.

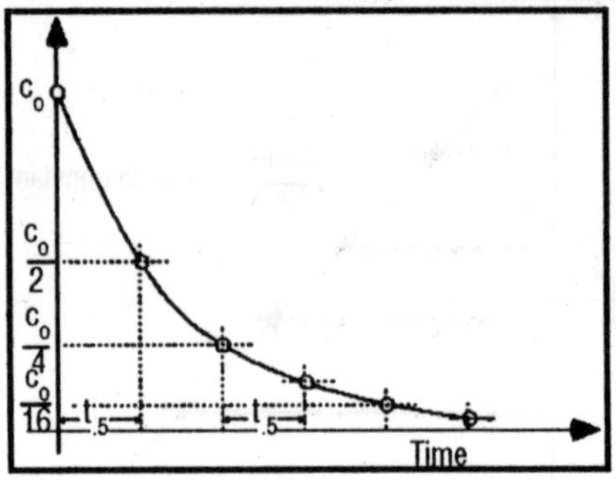

Fig. Concentration Independent Half life ($t_{.5}$) of First Order Reactions.

When discussing radioactive decay it is customary to call the "rate constant" (k) the "decay constant". It is furthermore customary to consider the time it takes to decrease the number of originally present species (normally called reactants) to 0.5 (by 50%). We then talk about the half-life. Accordingly, the above equation can be reformulated as:

$$k = \frac{2.3}{t_{1/2}} \times \log \frac{c_1}{\frac{c_1}{2}} = \frac{2.3}{t_{1/2}} \times \log 2$$

or

$$k = \frac{0.693}{t_{1/2}}$$

This means that from the half-life (t1/2) we may obtain the decay constant (rate constant in general). Vice versa, knowing the rate constant, we can calculate the time it takes to complete 50% of the reaction (or decay).

REACTION ORDER

Reactions of first order, such as the decay of H_2O_2 or radioactive decay, do not require molecular or atomic collisions - in principle they reflect inherent instability. A multitude of chemical reactions do involve collisions, however. For example, take the decay of HI:

$$2HI \rightarrow H_2 + I_2$$

The rate law for this reaction reflects the requirement of a collision in the concentration exponent:

$$-\frac{d[H]}{dt} = k[HI] = k[HI]^2$$

The reaction is referred to as a second order reaction which can readily be identified since upon integration the rate law yields:

$$\frac{1}{[HI]} = kt + const$$

or, more geneally:

$$\frac{1}{c} = kt + const$$

A plot of 1/c vs t will yield a straight line.

It is interesting to note that a determination of the reaction order from experimental data supplies insight to the details of how molecules and atoms react with each other.

Even though rate data involve measurements of gross numbers of molecules, their interpretation (through the reaction order) permits us to formulate the probable step (or steps) which individual molecules undergo.

TEMPERATURE DEPENDENCE OF REACTION RATES

Although a chemical reaction may be "thermodynamically" favored (which means the free energy of the system will be lowered as a result of the reaction), reaction may nevertheless not take place. Thus, H_2 and O_2 gases can exist in intimate contact over a considerable temperature range before reaction takes place. Nevertheless, a lighted match or a platinum powder catalyst can nucleate an immediate explosive reaction.

Another example with which we are familiar is glass: some glasses remain "supercooled liquids" for thousands of years at room temperature unless reheated to some definite temperature for at least some minimum period of time. Many different factors may influence the rate of a reaction - for example:

- Existing interatomic (or intermolecular) bonds must be broken.
- Atoms must be moved to and away from the reaction site.
- A new boundary is required wherever a new "phase" is to be nucleated.

All three steps cited require work or a supply of energy. Furthermore, all three steps are temperature dependent. At the beginning of this chapter we have stated: Rate = k(T) × function of concentration of reactants In other words: the reaction rate is proportional to the concentration of reactants. The proportionality constant k is a function of temperature k(T).

The temperature dependence of k (which represents the temperature dependence of the rate of a given reaction) can be studied by performing experiments (with constant concentrations) at different temperatures. From such experiments it can be seen that the value of the rate constant (k) is much greater at higher temperatures - the reactions proceed faster.

In 1899 Arrhenius showed that the rate constant of reactions increased in an exponential manner with T. By an empirical procedure he found that a plot of log k versus 1/T gives a linear relation. Remember: In all kinetic and thermodynamic calculations it is mandatory to use the "absolute temperature scale" in Kelvin (K) which is given by:

$$K = °C + 273.16$$

Thus 0K = –273.16 C, which corresponds to the thermodynamic absolute zero temperature. Linear plots of log k vs 1/T imply the relation:

$$\text{Ink} \propto \frac{1}{T}$$

$$k \propto^{const/1}$$

In view of later deductions this empirical relation can be conveniently written as:

$$k = Ae^{-E/RT}$$

where A is a proportionality constant sometimes called the reaction constant, R is the gas constant and equals 8.31 J/mole K, E is the activation energy in Joules/mole and e is the base of the natural (Naperian) system of logarithms. Since ln x = 2.3 log10x, we can rewrite the above equation:

$$\text{Ink} = \text{InA} - \frac{E}{RT}$$

or

$$\log k = \log A - \frac{E}{2.3 \times 8.3 \times T}$$

If the above equation is obeyed, a plot of log k versus 1/T will be a straight line from which the characteristic activation energy can be determined since the slope is –E/19.15. The reaction constant (log A) may in turn be evaluated by extrapolating the rate line to the point at which 1/T equals zero.

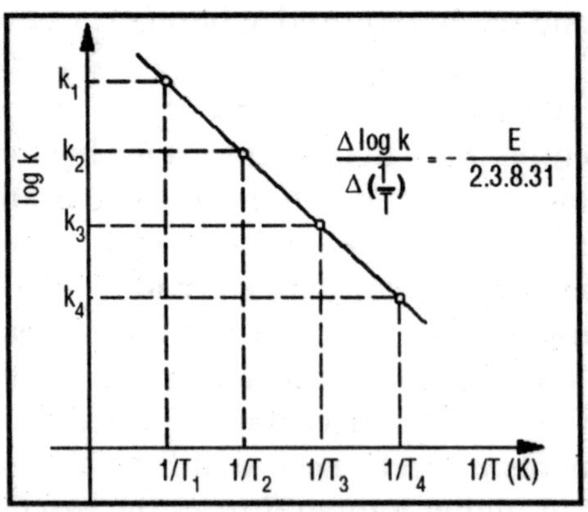

Fig. Arrhenius Plot Permitting the Determination of Activation Energies.

It is important to realise that many reactions involve a succession of steps, in which case the rate is controlled by the slowest step. The corresponding values of E and A must be determined by experimental methods.

THE ARRHENIUS THEORY

Arrhenius developed a primarily qualitative theory for molecular reactions which led to empirical expressions for the rate constant. Later theories (which are beyond the scope of 3.091) elaborate and make these original ideas more quantitative. The very simple (reversible) vapour-phase reaction of hydrogen and iodine to give hydrogen iodide can be used to illustrate the ideas of Arrhenius. The reaction

$$H_2 + I \rightarrow 2HI$$

apparently proceeds by a one-step (one collision) four-centre process such that the path of the reaction can be depicted as follows:

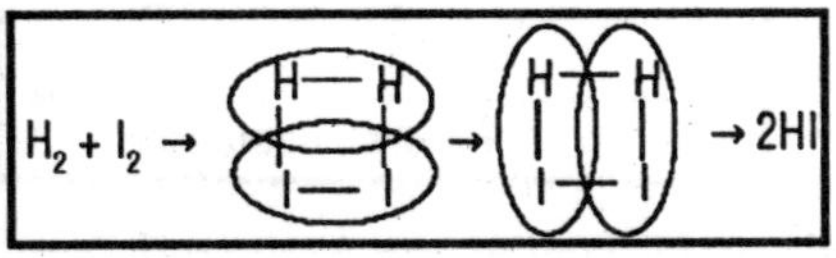

Fig. Vapour-Phase Reaction of Hydrogen and Iodine.

For more complicated systems, even when a mechanism has been postulated, it is not so easy to see how the electrons and atoms move around as the reaction proceeds.

However, even Arrhenius recognized that any reaction process can proceed first by means of the formation of some "high-energy species" (which we now call the "activated-complex") and secondly by the breakdown of this complex into products.

If the activated-complex is assumed to have an energy, Ea, greater than the reactants, then, in analogy to earlier considerations, the number of activated-complex molecules compared with the number of reactant molecules can be written in terms of the Boltzmann distribution as:

$$\frac{[\text{activated–complex molecules}]}{[\text{reactant}]} = e^{-E_a/RT}$$

The rate of reaction thus becomes proportional to the concentration of activated-complex molecules:

$$\text{Rate } \mu \text{ (activated-complex molecules)}$$

or

$$\text{Rate} = A \times e^{-E_a/RT} \times \text{Reactants]}$$

Making

$$k = Ae^{-E_a/RT}$$

we find

$$\text{Rate} = k\,[\text{reactants}]$$

Moreover, the theory says that the empirical constant, E_a, is to be interpreted as the energy of the activated-complex compared with that of the reactant molecules.

The idea of an activated-complex can be presented on a plot of the energy of

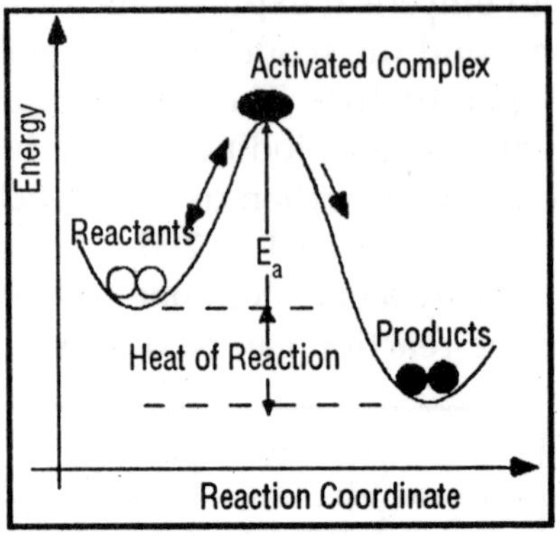

Fig. Schematic of a Reaction Via Activated Complex Formation.

the system as ordinate versus the reaction coordinate as abscissa. The reaction coordinate is not any single internuclear distance, but rather depends on all the internuclear distances that change as the reactant molecules are converted into product molecules.

In general it is impossible, and for the present purpose unnecessary, to give a quantitative description of the reaction coordinate. It consists of the transformation from reactants to products.

The Arrhenius theory leads to a considerable improvement in our understanding of the reaction process. It is, however, still a very qualitative theory in that it does not show how the pre-exponential factor A depends on the molecular properties of the reaction system, nor does it attempt to predict the value of Ea.

THE ACTIVATION ENERGY

The basic distribution of molecular (or atomic) energies at two different temperatures. It can be shown that $e^{-E_a/RT}$ is the fraction of molecules or atoms having an energy of Ea or greater.

The Arrhenius equation holds only if the interacting species have between them at least the certain critical energy Ea. Since the fraction having this energy or greater is $e^{-E_a/RT}$, the reaction rate is proportional to this quantity.

As the value of Ea increases, the energy requirement increases and it becomes more difficult for the molecules to acquire this energy. In contrast, e-Ea/RTincreases rapidly with increasing temperature (T).

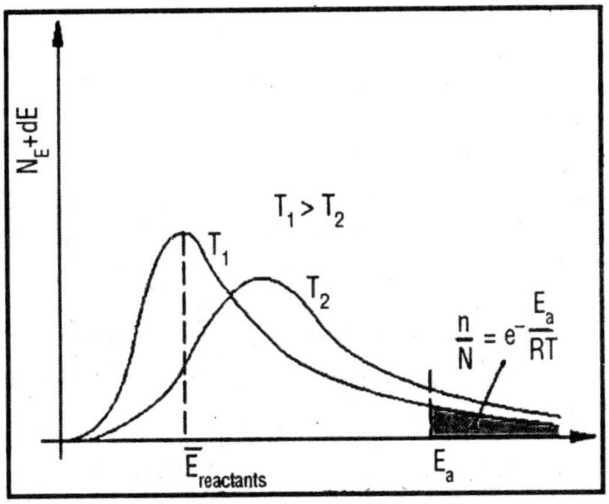

Fig. Maxwell Boltzmann Energy Distribution.

The change of a reaction rate with increasing temperature is usually much greater than expected from the corresponding increase in the average velocity of molecules and atoms. The average velocity of molecules and atoms is proportional to the square root of the absolute temperature.

Thus, if the temperature is raised 10σfrom 298σto 308?, the average velocity increase, (308/298)1/2, is but 2% whereas the rate of reaction increases by about 100%. From this we must conclude that the reaction rate is controlled not only by the number of collisions, but also by the activation energy.

We now realise that in order for a reaction to proceed we have to supply activation energy. We also know that for given conditions of concentration and T the rate of any reaction is inversely proportional to the energy of activation.

Since, particularly in industry, time is a major factor which can make a process economically feasible or unfeasible, considerable efforts have been put into accelerating reactions by means of catalysts. In catalytic reactions the "catalyst", which effectively lowers the required activation energy, is characterized by the following criteria:

- Unchanged chemically at the end of a reaction;
- Required in small amounts only;
- Catalytic action is frequently proportional to its surface area;
- Catalysis can be selective: if different reactions are possible, catalysis can enhance the rate of either one without affecting the alternate reaction.

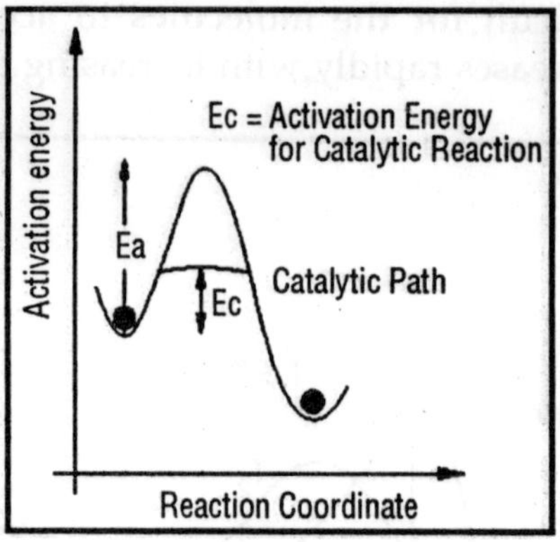

Fig. Effect of Catalysis on Activation Energy in Chemical Reaction.

For example:

$C_2H_5OH \rightarrow H_2C{=}CH_2 + H_2O$ on γ Alumina

(important in polymer chemistry)

(γAlumina: a cubic form of Al_2O_3)

$$C_2H_5OH \rightarrow CH_3-C\begin{smallmatrix}/\!\!/O \\ \backslash H\end{smallmatrix} + H_2$$

While the exact action of catalysis is still quite unclear, we do have some concrete information. The most frequently encountered case, heterogeneous catalysis (the presence of certain solids which increase gas reactions) is attributed to activation by adsorption.

From infrared studies we know that upon adsorption of compounds the bonds within the compound are weakened and reactions can subsequently occur with decreased activation energy.

Catalysis has certainly had considerable impact on our daily life. Until 1940 gasoline was exclusively made from crude oil. Also, possibly more important, nitrates and ammonia came primarily from Chile (Chile saltpeter). In the Fischer-Tropsch process, coal and steam are converted into gasoline hydrocarbons with the aid of Ni-Co catalysts.

The Haber-Bosch process, $N_2 + 3H_2 \rightarrow 2NH_3$ (ammonia) + 99 kJ, is likely the most important catalytic reaction practiced: From the "Le Chatelier" principle (if in a reaction the number of molecules present decreases, the reaction can be accelerated by applying increased pressures) we know that high pressures will favour this synthesis, but we also know that high temperature will favour the reverse reaction, i.e., decay of ammonia; at high temperatures complex molecular structures tend to decay to more elemental, basic species.

Catalysis (Fe, Al_2O_3 + K_2O in solid form) makes it possible at 400°C and at about 600 atm to convert about 60% of the gas mixture $N_2 + 3H_2$ to

NH_3. Nitrates (important as fertilizers and explosives) are predominantly produced by catalytic oxidation.

$4NH_3 + 5O_2 \rightarrow 4NO + 6H_2O$ over Pt at 900°C

$2NO + O_2 \rightarrow 2NO_2$

$3NO_2 + H_2O \rightarrow 2HNO_3$ (Nitric Acid) + NO

Other examples are:

- $SO_2 + 1/2\ O_2 \rightarrow SO_3$ over V_2O_5
- Synthetic rubber from butadiene and styrene
- Methanol from $CO + 2H_2 \rightarrow CH_3OH$ over $ZnO + Cr_2O_3$
- Formaldehyde from $CH_3OH \rightarrow CH{\overset{/\!/O}{\underset{\backslash H}{}}}$ over Cu (important for plastics)
- In nature we have a large number of catalysts in the form of enzymes.

Chapter 10

Diffusion

At any temperature different from absolute zero all atoms, irrespective of their state of aggregation (gaseous, liquid or solid), are constantly in motion. Since the movement of particles is associated with collisions, the path of a single particle is a zigzag one. However, an aggregation of "diffusing" particles has an observable drift from places of higher to places of lower concentration. For this reason diffusion is known as a transport phenomenon.

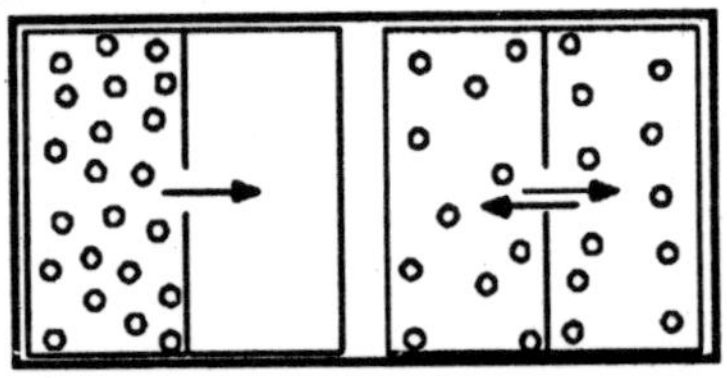

Fig. Mass Transport, Diffusion as a Consequence of Existing Spacial Differences in Concentration.

In each diffusion reaction (heat flow, for example, is also a diffusion process), the flux (of matter, heat, electricity, etc.) follows the general relation:

Flux = (conductivity) × (driving force)

In the case of atomic or molecular diffusion, the "conductivity" is referred to as the diffusivity or the diffusion constant, and is represented by the symbol D.

We realise from the above considerations that this diffusion constant (D) reflects the mobility of the diffusing species in the given environment and accordingly assumes larger values in gases, smaller ones in liquids, and extremely small ones in solids.

The "driving force" for many types of diffusion is the existence of a concentration gradient. The term "gradient" describes the variation of a given property as a function of distance in the x-direction. If a material exhibits a linear variation of concentration with distance in the x-direction, we speak of a constant concentration gradient in the x-direction. The gradient itself is the rate of change of the concentration with distance (dc/dx), which is the same as the slope of a graph of concentration vs. position (Δ c/Δ x).

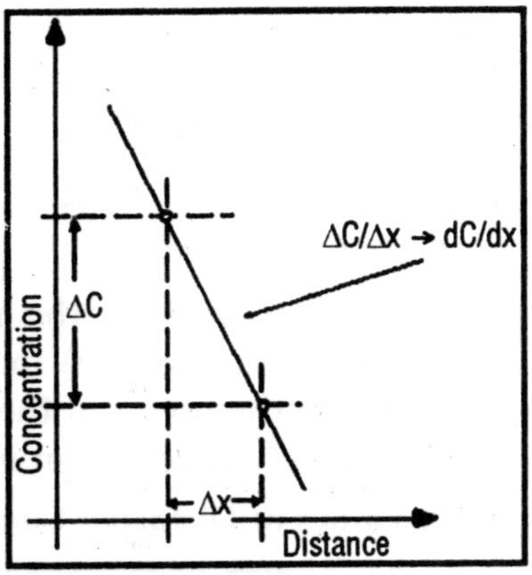

Fig. Concentration Gradient (Constant) in the x Direction.

Steady State and Nonsteady Diffusion

Diffusion processes may be divided into two types:

(a) Steady state

(b) Nonsteady state.

Steady state diffusion takes place at a constant rate - that is, once the process starts the number of atoms (or moles) crossing a given interface (the flux) is constant with time. This means that throughout the system dc/dx = constant and dc/dt = 0. Nonsteady state diffusion is a time dependent process in which the rate of diffusion is a function of time. Thus dc/dx varies with time and dc/dt = 0.

Both types of diffusion are described quantitatively by Fick's laws of diffusion. The first law concerns both steady state and nonsteady state diffusion, while the second law deals only with nonsteady state diffusion.

STEADY STATE DIFFUSION (FICK'S FIRST LAW)

On the basis of the above considerations, Fick's First Law may be formulated as:

$$J = -D\frac{dc}{dx}$$

In words: The diffusive flux isproportional to the existing concentration gradient.

The negative sign in this relationship indicates that particle flow occurs in a "down" gradient direction, i.e. from regions of higher to regions of lower concentration. The flux J can be given in units of atoms/

cm^2s, moles/cm^2s, or equivalents. Correspondingly, the diffusivity (D) will assume the dimensions cm^2/s, as can be seen from a dimensional analysis:

$$J = \left(\frac{\text{Moles}}{\text{cm}^2/\text{s}}\right) = -D\left(\frac{dc}{dx}\right)\left(\frac{\text{Moles.cm}^{-3}}{\text{cm}}\right)$$

$$\text{Thus}: D = \text{cm}^2/\text{s}$$

Like chemical reactions, diffusion is a thermally activated process and the temperature dependence of diffusion appears in the diffusivity

$$D = D_o e^{-Ea/RT}$$

(a) where D_o (the equivalent of A in the previously discussed temperature dependence of the rate constant) includes such factors as the jump distance, the vibrational frequency of the diffusing species and so on. Selected values of D, Do and Ea are given in Table.

Table. Selected Values of Diffusion Constants (D).

Diffusing Substance	Solvent	T (° C)	D ($cm^2.s^{-1}$)
Au	Cu	400	5×10^{-13}
Cu (Self-Diffusing)	(Cu)	650	3.2×10^{-12}
C	Fe (FCC)	950	10^{-7}
Methanol	H_2O	18	1.4×10^{-5}
O_2	Air	0	0.178
H_2	Air	0	0.611

Table. Selected Values of Do and Ea for Diffusion Systems.

Solute	Solvent (host structure)	Do, cm^2s	Ea , kJoules/mole
1. Carbon	fcc iron	0.2100	142
2. Carbon	bcc iron	0.0079	76
3. Iron	fcc iron	0.5800	285
4. Iron bcc	iron	5.8000	251
5. Nickel	fcc iron	0.5000	276
6. Manganese	fcc iron	0.3500	283
7. Zinc	copper	0.0330	159
8. Copper	aluminum	2.0000	142
9. Copper	copper	11.0000	240
10. Silver	silver	0.7200	188

A typical application of Fick's first law: Determine the rate at which helium (He), held at 5 atm and 200°C in a Pyrex glass bulb of 50 cm diameter and a wall thickness (x) of 0.1 cm, diffuses through the Pyrex to the outside. Assume that the pressure outside the tube at all times remains negligible. (For the diffusion of gases it is

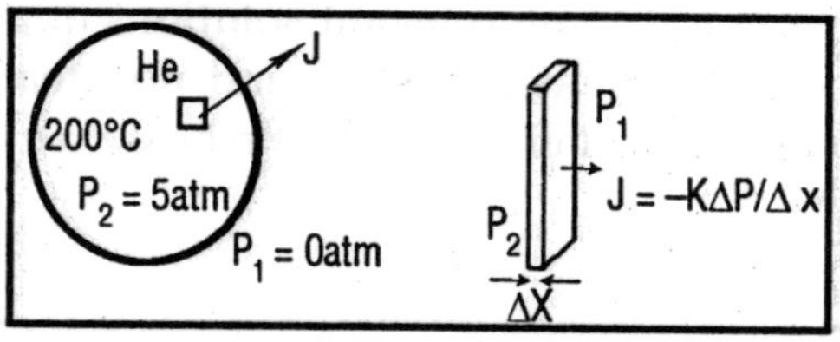

Fig. Conditions for Outdiffusion of He from a Glass Bulb.

customary, although not necessary, to replace the diffusion constant D with the Permeation constant K, normally given in units of $cm^2/s.atm$. Using the gas laws, K is readily converted to D if so desired.)

In the present system

$K = 1 \times 10^{-9}\ cm^2/s.atm$

We can now set up the diffusion equation:

$$J = -k\left(\frac{dp}{dx}\right)$$

We may now formally separate the variables and integrate:

$Jdx = -Kdp$

$$\int_{x=0}^{x=0.1} Jdx = -\int_{P_2=5}^{P_1 0} kdp$$

$$J = K\frac{5.0}{0.1}$$

We can forego the integration since $(dP/dx) = (\Delta P/\Delta x)$ and we may immediately write:

$$J = (\text{total}) = -K\frac{\Delta P}{\Delta X} \times A = 5 \times 10^{-8} \times 7.9 \times 10^{3}$$

$$J = 3.9 \times 10^{-4}$$

The units of the flux may be obtained from a dimensional analysis:

$$J = -K\frac{\Delta P}{\Delta X} \times A = \frac{cm^2}{s \cdot atm}\frac{atm}{cm}\frac{cm^2}{1} = \frac{cm^3}{s}$$

The total flux is $3.9 \times 10^{-4}\ cm^3/s$ (with the gas volume given for 1 atm). If the total gas flow by diffusion were to be determined for a specified time interval, the volume would be multiplied by the indicated time.

NONSTEADY STATE DIFFUSION (FICK'S SECOND LAW)

The quantitative treatment of nonsteady state diffusion processes is formulated as a partial differential equation. It is beyond the scope of 3.091 to treat the equations in detail but we can consider the second law qualitatively and examine some relevant solutions quantitatively.

The difference between steady state and nonsteady state diffusion conditions can readily be visualized. In the first case we have, for example, the diffusion of gas

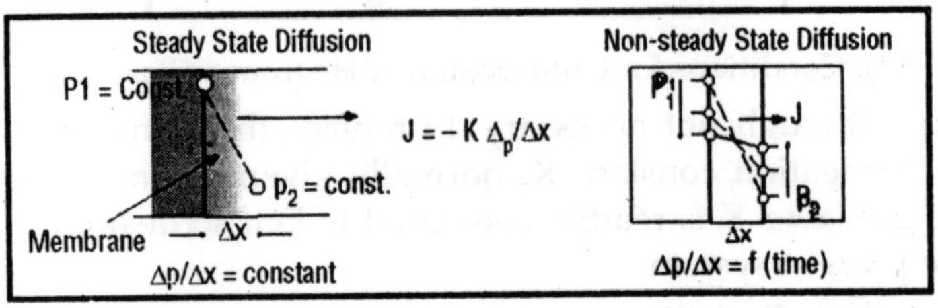

Fig. Steady State and Non-steady State Diffusion.

from an infinite volume (P_1 const) through a membrane into an infinite volume (P2 const). The pressure gradient across the membrane remains constant as does the diffusive flux. In the second case we deal with diffusion from a finite volume through a membrane into a finite volume. The pressures in the reservoirs involved change with time as does, consequently, the pressure gradient across the membrane.

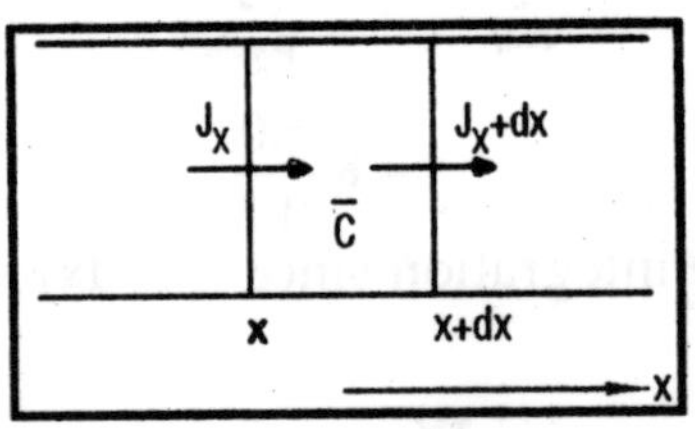

Fig. Ressure Gradient.

Consider a volume element (between x and x + dx of unit cross sectional area) of a membrane separating two finite volumes involved in a diffusion system. The flux of a given material into the volume element minus the flux out of the volume element equals the rate of accumulation of the material into this volume element:

$$J_x - J_{x+dx} = \frac{\overline{\partial c}}{\partial t} dx$$

[$\bar{c}$ is the average concentration in the volume element and $\bar{c}\,dx$ is the total amount of the diffusing material in the element at time (t).]

Using a Taylor series we can expand Jx+dx about x and obtain:

$$J_{x+dx} + J_x + \frac{\partial J_x}{\partial x}dx + \frac{\partial^2 J_x}{\partial x^2}\frac{dx^2}{2} + \ldots$$

Accordingly, as dx → 0:

$$\frac{\partial}{\partial x} = \left(D\frac{\partial c}{\partial x}\right) = \frac{\partial c}{\partial t}$$

and if D does not vary with x (which is normally the case) we have the formulation of Fick's Second Law:

$$\frac{\partial c}{\partial t} = D\frac{\partial^2 c_x}{\partial x^2} \text{(Fick's Second Law)}$$

In physical terms this relationship states that the rate of compositional change is proportional to the "rate of change" of the concentration gradient rather than to the concentration gradient itself.

The solutions to Fick's second law depend on the boundary conditions imposed by the particular problem of interest. As an example, let us consider the following problem (encountered in many solid state processes):

A frequently encountered situation is the diffusion of a component 2 into an infinite region of a material 1 [planar diffusion of doping elements into semiconductors

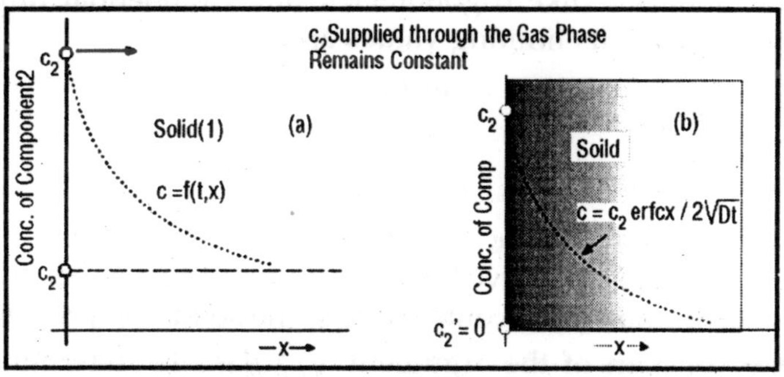

Fig. Diffusion at Constant Surface Concentration; Bulk Concentration of Component 2 at Start of Diffusion is c2' (in a) and is Zero (in b).

for the generation of junction devices (p-n junctions, junction transistors)]. The boundary conditions are: the concentration of component at the surface of the solid phase (x=0) remains constant at c_2 and the concentration of component in the solid prior to diffusion is uniformly.

Under these boundary conditions the solution to Fick's second law assumes the form:

$$\frac{c_2 - c}{c_2 - c_2'} \operatorname{erf}\left(\frac{x}{2\sqrt{Dt}}\right)$$

If no component is originally in the solid matrix, the above solution is simpler:

$$\frac{c_2 - c}{c_2} \operatorname{erf}\left(\frac{x}{2\sqrt{Dt}}\right)$$

$$1 - \frac{c}{c_2} = \operatorname{erf}\left(\frac{x}{2\sqrt{Dt}}\right)$$

$$\frac{c}{c_2} = 1 - \operatorname{erf}\left(\frac{x}{2\sqrt{Dt}}\right)$$

$$\frac{c}{c_2} = \operatorname{erfc}\left(\frac{x}{2\sqrt{Dt}}\right)$$

$$c = c_2 \operatorname{erfc}\left(\frac{x}{2\sqrt{Dt}}\right)$$

An analysis shows that the last form of the solution to Fick's law relates the concentration (c) at any position (x) (depth of penetration into the solid matrix) and time (t) to the surface concentration (c_2) and the diffusion constant (D). The terms erf and erfc stand for error function and complementary error function respectively - it is the Gaussian error function as tabulated (like trigonometric and exponential functions) in mathematical tables. Its limiting values are:

$$\operatorname{erf}(0) = 0$$

$$\operatorname{erf}(\infty) = 1$$

$$\operatorname{erf}(-\infty) = -1$$

And for the complementary error function: erfc = (1 – erf)

Another look at the above solution to the diffusion equation shows that the concentration (c) of component in the solid is expressed in terms of the error function of the argument $x/2\sqrt{Dt}$. To determine at what depth a particular concentration (c) of (2) will appear, we substitute this concentration for (c) and obtain:

$$\frac{c^*}{c_2} = \operatorname{erfc}\left(\frac{x}{2\sqrt{Dt}}\right) = K$$

Since the error function is a constant, its argument must also be a constant:

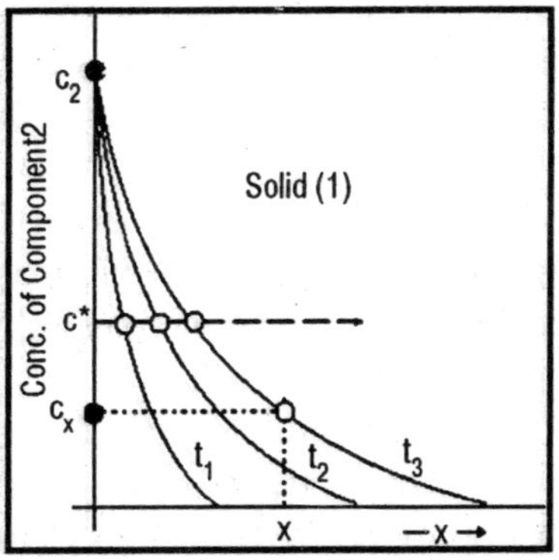

Fig. Advance of the Concentration Front (c) as a Function of Trime.

$$\frac{x}{2\sqrt{Dt}} = K'$$

Therefore, under the given boundary conditions:

$$x = K'2\sqrt{Dt}$$

$$x = K'\sqrt{Dt}$$

$$x \propto \sqrt{t}$$

The depth of penetration of a specified concentration is found to be proportional to the square root of the time of diffusion.

SELF-DIFFUSION

As previously indicated, the thermal motion of atoms in a lattice is a random process and as such will lead to local displacements of individual atoms. This random movement of atoms within a lattice (self-diffusion), which is not associated with any existing concentration gradients, can be readily demonstrated with the aid of "radioactive elements". For example, nickel appears in nature in the form of several "stable isotopes":

$$Ni_{28}^{58}, Ni_{28}^{60}, Ni_{28}^{62} \text{ and } Ni_{28}^{64}$$

If Ni_{28}^{58}, is irradiated with neutrons in a nuclear reactor, it will capture a neutron and

become Ni_{28}^{59}, which is radioactive (a radio-isotope).

$$Ni_{28}^{58} + n \rightarrow Ni_{28}^{59} \rightarrow Ni_{27}^{59} + \gamma + \beta$$

Nickel 59 is characterized by its instability which leads to the emission of β and γradiation, with a half-life of 8 × 10^4 years. Since this radiation can be measured by appropriate radiation detectors, it is possible to use Ni_{59} as a "tracer element" for studies of self-diffusion.

The radioactive nickel (which is identical to ordinary nickel with the exception of its radioactive properties) is electroplated onto normal nickel. This specimen is subsequently placed into a furnace and heated up to close to its melting point for an extended period of time. After removing the specimen, it is sectioned into slices parallel to the surface which contained the radioactive tracer element.

With the aid of a radiation detector it can now be shown that the Ni_{59}, which originally was only at the surface, has diffused into the bulk material while simultaneously some bulk nickel has counter-diffused in the other direction. If this sample is heat-treated for a much longer time, sectioning and counting will reveal a completely uniform distribution of the radio-tracer element.

It can thus be shown that self-diffusion does occur in solids, and quantitative measurements with tracer elements even permit the determination of self-diffusion coefficients.

DIFFUSION MECHANISMS

The diffusion process in interstitial solid solutions, like that of carbon in iron, can readily be understood as a result of considerable differences in atomic diameters. However, the fact that Au diffuses faster in Pb than NaCl diffuses in water at 15°C cannot be readily explained. The magnitude of the observed activation energy indicates that a mechanism whereby atoms simply change places with each other has to be excluded. More reasonable mechanisms were suggested by Frenkel and Schottky. They proposed the existence of point defects (vacancies) in crystals which provide a mechanism by which atoms can move (diffuse) within a crystal.

The concentration of such vacancies, as you recall, can be calculated from simple statistical calculations. In most solids we are not dealing with single crystals but rather with polycrystalline materials which contain a large number of grain boundaries (internal surfaces). As expected, the rate of diffusion along grain boundaries is much higher than that for volume diffusion ($D_{volume} < D_{g\text{-}boundary}$). FInally, surface diffusion, which takes place on all external surfaces, is even higher ($D_{volume} < D_{g\text{-}boundary} < D_{surface}$). The respective activation energies for diffusion are:

E_a surface < E_a grain boundary < E_a volume

Diffusion in Non-Metals

In non-metallic systems diffusion takes place by the same mechanisms as in metallic systems. Oxygen, for example, diffuses through many oxides by vacancy migration. In crystalline oxides and in silicate glasses as well, it is found that oxygen diffuses much more rapidly than the metallic ion. In glasses containing alkali atoms (Na^+, K^+), the respective rates of diffusion are:

$$D_{alkali} > D_{oxygen} > D_{silicon}$$

corresponding to differences in bonding strengths. In polymer materials diffusion requires the motion of large molecules since intramolecular bonding is much stronger than intermolecular bonding. This fact explains that the diffusion rates in such materials are relatively small.

Gaseous Diffusion in Solids

Some gases, like hydrogen and helium, diffuse through some metals with ease even at room temperature. Helium, for example, will diffuse through quartz and steel and limits the ultimate vacuum obtainable in ultra-high vacuum systems.

Hydrogen similarly diffuses readily through Ni at elevated temperatures. H_2 also diffuses at high rates through palladium - a phenomenon which is used extensively for hydrogen purification since that material is impervious to other gases.

PHASE EQUILIBRIA AND PHASE DIAGRAMS

Phase diagrams are one of the most important sources of information concerning the behaviour of elements, compounds and solutions.

They provide us with the knowledge of phase composition and phase stability as a function of temperature (T), pressure (P) and composition (C). Furthermore, they permit us to study and control important processes such as phase separation, solidification, sintering, purification, growth and doping of single crystals for technological and other applications.

Although phase diagrams provide information about systems at equilibrium, they can also assist in predicting phase relations, compositional changes and structures in systems not at equilibrium.

GASES, LIQUIDS AND SOLIDS

Any material (elemental or compound) can exist as a gas, a liquid or a solid, depending on the relative magnitude of the attractive interatomic or intermolecular forces vs the disruptive thermal forces. It is thus clear that the stability (existence) of the different states of aggregation, which are referred to as phases, is a function of temperature and pressure (since with increased pressure the atoms, for exampled of a gas phase, are closer spaced and thus subject to increased interatomic attraction).

In general terms, a "phase" is a homogeneous, physically distinct, mechanically separable portion of a material with a given chemical composition. To illustrate this definition, let us look at a few examples of common multi-phase systems. Ice cubes in water constitute a two-phase system (ice and liquid water), unless we include the vapour above the glass in our system, which would make it a three-phase system. A mixture of oil and water would also be a two-phase system. Just as oil and water represent two distinct liquid phases, two regions of a solid with distinctly different composition or structure separated by boundaries represent two solid phases.

If we look at a one-component system, such as liquid water, we recognize that because of the energy distribution of the water molecules, some water molecules will always possess sufficient energy to overcome the attractive forces on the surface of H_2O and enter into the gas phase. If thermal energy is continuously supplied to a liquid in an open container, the supply of high energy molecules (which leave the liquid phase) is replenished and the temperature remains constant – otherwise the loss of high energy molecules will lower the temperature of the system.

The total quantity of heat necessary to completely "vaporize" one mole of a liquid at its boiling point is called its molar heat of vaporization, designated by Δ HV. Similarly, the heat required to completely melt one mole of a solid (the heat required to break the bonds established in the solid phase) is called the (latent) heat of fusion (Δ HV).

Visualize a liquid in a sealed container with some space above the liquid surface. Again, some of the most energetic liquid molecules will leave the liquid phase and form a "gas phase" above the liquid.

Since gas molecules will thus accumulate in the gas phase (at a constant temperature), it is inevitable that as a result of collisions in the gas phase some molecules will re-enter the liquid phase and a situation will be established whereby the rate of evaporation will equal the rate of condensation – i.e., a dynamic equilibrium between the liquid and gas phase will exist. The established pressure in the gas phase is referred to as the equilibrium vapour pressure, which is normally significantly less for solids than for liquids.

For obvious reasons it is desirable to know for any given material the conditions (P, T) under which the solid state, the liquid state and the gaseous state are stable, as well as the conditions under which the solid and liquid phases may coexist. These conditions are graphically presented in equilibrium phase diagrams, which can be experimentally determined.

THE ONE-COMPONENT PHASE DIAGRAM

Figure illustrates the temperatures and pressures at which water can exist as a solid, liquid or vapour. The curves represent the points at which two of the phases coexist in equilibrium. At the pcint Tt vapour, liquid and solid coexist in equilibrium. In the fields of the diagram (phase fields) only one phase exists. Although a diagram of this kind delineates the boundaries of the phase fields, it does not indicate the quantity of any phase present.

It is of interest to consider the slope of the liquid/solid phase line of the H_2O phase diagram. It can readily be seen that if ice – say at –2°C – is subjected to high pressures, it will transform to liquid H_2O. (An ice skater will skate not on ice, but on water.)

This particular pressure sensitivity (reflected in the slope of the solid/liquid phase line) is characteristic for materials which have a higher coordination number in the liquid than in the solid phase (H_2O, Bi, Si, Ge). Metals, for example have an opposite slope of the solid/liquid phase line, and the liquid phase will condense under pressure to a solid phase.

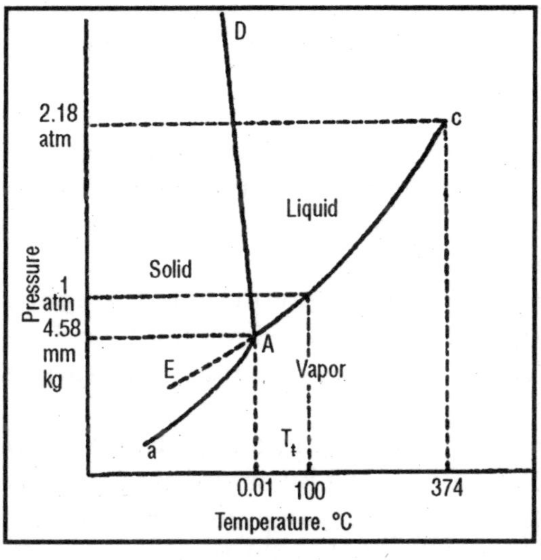

Fig. Pressure/Temperature Diagram for Water. (Not Drawn to Scale.)

PHASE RULE AND EQUILIBRIUM

The phase rule, also known as the Gibbs phase rule, relates the number of components and the number of degrees of freedom in a system at equilibrium by the formula

$$F = C - P + 2$$

where F equals the number of degrees of freedom or the number of independent variables, C equals the number of components in a system in equilibrium and P equals the number of phases. The digit 2 stands for the two variables, temperature and pressure.

The number of degrees of freedom of a system is the number of variables that may be changed independently without causing the appearance of a new phase or disappearance of an existing phase. The number of chemical constituents that must be specified in order to describe the composition of each phase present. For example, in the reaction involving the decomposition of calcium carbonate on heating, there are three phases – two solid phases and one gaseous phase.

$$CaCO_3\ (s) \leftrightarrows CaO\ (s) + CO_2\ (g)$$

There are also three different chemical constituents, but the number of components is only two because any two constituents completely define the system in equilibrium. Any third constituent may be determined if the concentration of the other two is known.

Substituting into the phase rule we can see that the system is univariant, since $F = C - P + 2 = 2 - 3 + 2 = 1$. Therefore only one variable, either temperature or pressure, can be changed independently. (The number of components is not always easy to determine at first glance, and it may require careful examination of the physical conditions of the system at equilibrium.)

The phase rule applies to dynamic and reversible processes where a system is heterogeneous and in equilibrium and where the only external variables are temperature, pressure and concentration. For one-component systems the maximum number of variables to be considered is two – pressure and temperature.

Such systems can easily be represented graphically by ordinary rectangular coordinates. For two-component (or binary) systems the maximum number of variables is three – pressure, temperature and concentration.

Only one concentration is required to define the composition since the second component is found by subtracting from unity. A graphical representation of such a system requires a three-dimensional diagram. This, however, is not well suited to illustration and consequently separate two-coordinate diagrams, such as pressure vs temperature, pressure vs composition and temperature vs composition, are mostly used. Solid/liquid systems are usually investigated at constant pressure, and thus only two variables need to be considered – the vapour pressure for such systems can be neglected.

This is called a condensed system and finds considerable application in studying phase equilibria in various engineering materials. A condensed system will be represented by the following modified phase rule equation:

$$F = C - P + 1$$

where all symbols are the same as before, but (because of a constant pressure) the digit 2 is replaced by the digit 1, which stands for temperature as variable. The graphical representation of a solid/liquid binary system can be simplified by representing it on ordinary rectangular coordinates: temperature vs concentration or composition.

H vs T PHASE DIAGRAM

With the aid of a suitable calorimeter and energy reservoir, it is possible to measure the heat required to melt and evaporate a pure substance like ice. The experimental data obtainable for a mole of ice is shown schematically in figure. As heat is added to the solid, the temperature rises along line "a" until the temperature of fusion (T_f) is reached. The amount of heat absorbed per mole during melting is represented by the length of line "b", or ΔH_F. The amount of heat absorbed per mole during evaporation at the boiling point is represented by line "d". The reciprocal of the slope of line "a", (dH/dT), is the heat required to change the temperature of one mole of substance (at constant pressure) by 1CF. (dH/dT) is the molar heat capacity of a material, referred to as "Cp". As the reciprocal of line "a" is Cp (solid), the reciprocals of lines "c" and "e" are C_p (liquid) and C_p (vapour) respectively.

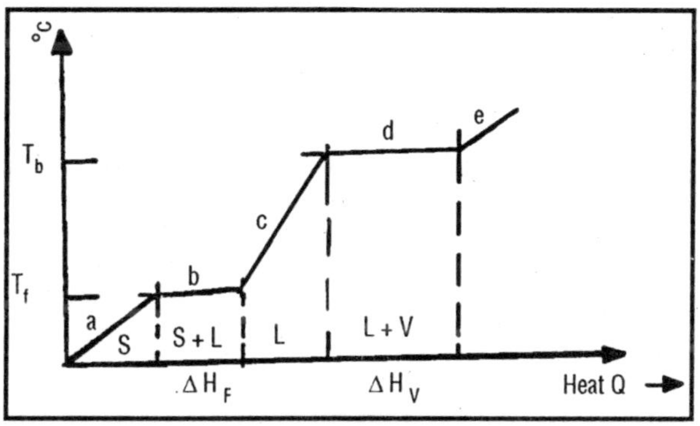

Fig. H vs T Diagram for Pure H2O. (Not to Scale.)

From a thermodynamic standpoint, it is important to realise that figure illustrates the energy changes that occur in the system during heating. Actual quantitative measurements show that 5.98 kJ of heat are absorbed at the melting point (latent heat of fusion) and 40.5 kJ per mole (latent heat of evaporation) at the boiling point.

The latent heats of fusion and evaporation are unique characteristics of all pure substances. Substances like Fe, Co, Ti and others, which are allotropic (exhibit different structures at different temperatures), also exhibit latent heats of transformation as they change from one solid state crystal modification to another.

ENERGY CHANGES

When heat is added from the surroundings to a material system (as described above), the energy of the system changes. Likewise, if work is done on the surroundings by the material system, its energy changes. The difference in energy (Δ E) that the system experiences must be the difference between the heat absorbed (Q) by the system and the work (W) done on the surroundings. The energy change may therefore be written as:

$$\Delta E = Q - W$$

If heat is liberated by the system, the sign of Q is negative and work done is positive. Q and W depend on the direction of change, but Δ E does not. The above relation is one way of representing the First Law of Thermodynamics which states that the energy of a system and its surroundings is always conserved while a change in energy of the system takes place. The energy change, Δ E, for a process is independent of the path taken in going from the initial to the final state.

In the laboratory most reactions and phase changes are studied at constant pressure.

The work is then done solely by the pressure (P), acting through the volume change, Δ V.

$$W = PDV \text{ and } DP = 0$$

Hence:

$$Q = \Delta E + P\Delta V$$

Since the heat content of a system, or the enthalpy H, is defined by:

$$H = E + PV$$

$$DH = DE + PDV$$

so that:

$$DH = Q - W + PDV$$

or

$$DH = Q$$

Reactions in which Δ H is negative are called exothermic since they liberate heat, whereas endothermic reactions absorb heat. Fusion is an endothermic process, but the reverse reaction, crystallization, is an exothermic one.

ENTROPY AND FREE ENERGY

When a gas condenses to form a liquid and a liquid freezes to form a crystalline solid, the degree of internal order increases. Likewise, atomic vibrations decrease to zero when a perfect crystal is cooled to 0°K. Since the term entropy, designated by S, is considered a measure of the degree of disorder of a system, a perfect crystal at 0°K has zero entropy.

The product of the absolute temperature, T, and the change in entropy, Δ S, is called the entropy factor, TΔ S. This product has the same units (Joules/mole) as the change in enthalpy, Δ H, of a system. At constant pressure, P, the two energy changes are related to one another by the Gibbs free energy relation:

$$D F = D H - TD S$$

where

$$F = H - TS$$

The natural tendency exhibited by all materials systems is to change from one of higher to one of lower free energy. Materials systems also tend to assume a state of greater disorder whereby the entropy factor TΔ S is increased. The free energy change, Δ F, expresses the balance between the two opposing tendencies, the change in heat content (Δ H) and the change in the entropy factor (TΔ S). If a system at constant pressure is in an equilibrium state, such as ice and water at 0°C, for example, at atmospheric pressure it cannot reach a lower energy state. At equilibrium in the ice-water system, the opposing tendencies, Δ H and TΔ S, equal one another so that Δ F = 0. At the fusion temperature, T_F:

$$\Delta S_F = \frac{\Delta H_F}{T_F}$$

Similarly, at the boiling point:

$$\Delta S_V = \frac{\Delta H_V}{T_V}$$

Thus melting or evaporation only proceed if energy is supplied to the system from the surroundings. The entropy of a pure substance at constant pressure increases with temperature according to the expression:

$$\Delta S = \frac{C_P \Delta T}{T} (\text{Since} : \Delta H = C_P \Delta T)$$

where Cp is the heat capacity at constant pressure, Δ Cp, Δ H, T and Δ T are all measurable quantities from which Δ S and Δ F can be calculated.

F vs T

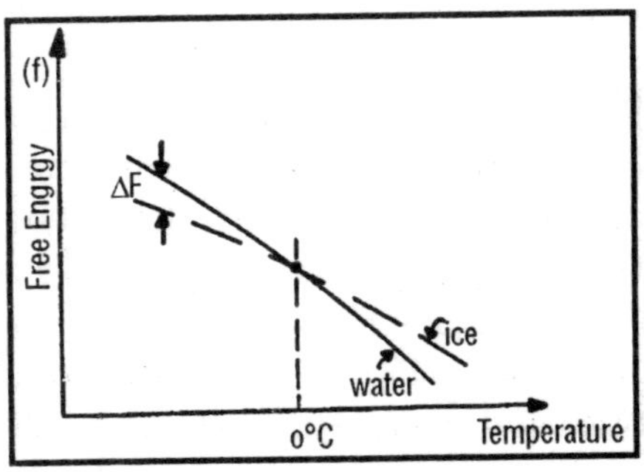

Fig. Free Energy is a Function of Temperature for Ice and Water.

Any system can change spontaneously if the accompanying free energy change is negative. This may be shown graphically by making use of F vs T curves such as those shown in figure.

The general decrease in free energy of all the phases with increasing temperature is the result of the increasing dominance of the temperature-entropy term. The increasingly negative slope for phases which are stable at increasingly higher temperatures is the result of the greater entropy of these phases.

PHASE DIAGRAMS (TWO-COMPONENT SYSTEMS)

Solid Solutions: A solution can be defined as a homogeneous mixture in which the atoms or molecules of one substance are dispersed at random into another substance. If this definition is applied to solids, we have a solid solution. The term "solid solution" is used just as "liquid solution" is used because the solute and solvent atoms (applying the term solvent to the element in excess) are arranged at random. The properties and composition of a solid solution are, however, uniform as long as it is not examined at the atomic or molecular level.

Solid solutions in alloy systems may be of two kinds: substitutional and interstitial. A substitutional solid solution results when the solute atoms take up the positions of the solvent metal in the crystal lattice. Solid solubility is governed by the comparative size of the atoms of the two elements, their structure and the difference in electronegativity. If the atomic radii of a solvent and solute differ by more than 15% of the radius of the solvent, the range of solubility is very small.

When the atomic radii of two elements are equal or differ by less than 15% in size and when they have the same number of valency electrons, substitution of one kind of atom for another may occur with no distortion or negligible distortion of the crystal lattice, resulting in a series of homogeneous solid solutions. For an unlimited solubility in the solid state, the radii of the two elements must not differ by more than 8% and both the solute and the solvent elements must have the same crystal structure.

In addition to the atomic size factor, the solid solution is also greatly affected by the electronegativity of elements and by the relative valency factor. The greater the difference between electronegativities, the greater is the tendency to form compounds and the smaller is the solid solubility. Regarding valency effect, a metal of lower valency is more likely to dissolve a metal of higher valency. Solubility usually increases with increasing temperature and decreases with decreasing temperature. This causes precipitation within a homogeneous solid solution phase, resulting in hardening effect of an alloy. When ionic solids are considered, the valency of ions is a very important factor. An interstitial solid solution results when the solute atoms are small enough to fit into the interstices of the metal lattice. The elements that can form interstitial solid solutions with transition metals are hydrogen, carbon, nitrogen and boron.

EQUILIBRIUM PHASE DIAGRAMS OF TWO-COMPONENT SYSTEMS

To construct an equilibrium phase diagram of a binary system, it is a necessary and sufficient condition that the boundaries of one-phase regions be known. In other words, the equilibrium diagram is a plot of solubility relations between components of the system. It shows the number and composition of phases present in any system under equilibrium conditions at any given temperature. Construction of the diagram is often based on solubility limits determined by thermal analysis – i.e., using cooling curves. Changes in volume, electrical conductivity, crystal structure and dimensions can also be used in constructing phase diagrams.

The solubility of two-component (or binary) systems can range from essential insolubility to complete solubility in both liquid and solid states, as mentioned above. Water and oil, for example, are substantially insoluble in each other while water and alcohol are completely intersoluble. Let us visualize an experiment on the water-ether system in which a series of mixtures of water and ether in various proportions is placed in test tubes.

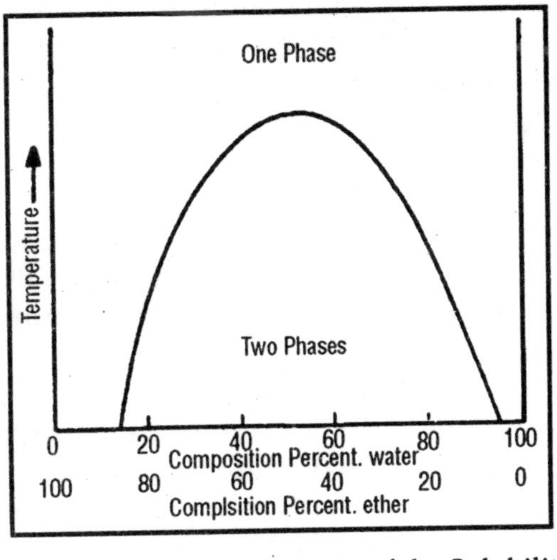

Fig. Schematic Representation of the Solubilities of Ether and Water in Each other.

After shaking the test tubes vigorously and allowing the mixtures to settle, we find present in them only one phase of a few per cent of ether in water or water in ether, whereas for fairly large percentages of either one in the other there are two phases. These two phases separate into layers, the upper layer being ether saturated with water and the lower layers being water saturated with ether.

After sufficiently increasing the temperature, we find, regardless of the proportions of ether and water, that the two phases become one. If we plot solubility limit with temperature as ordinate and composition as abscissa, we have an isobaric [constant pressure (atmospheric in this case)] phase diagram. This system exhibits a solubility gap.

COOLING CURVES

Equilibrium phase diagrams, also called constitutional diagrams, are usually constructed from the data obtained from cooling curves. These are secured by plotting the measured temperatures at equal time intervals during the cooling period of a melt to a solid. The main types of cooling curves are shown in figure.

The cooling curve shown in figure is that of a pure molten metal or a pure compound. The cooling proceeds uniformly along curve AB at a decreasing rate until point B is reached – when the first crystals being to form. As freezing continues the latent heat of fusion is liberated in such amounts that the temperature remains constant from B to C until the whole mass has entirely solidified.

Period BC is known as the "horizontal thermal arrest". Further cooling from point C will cause the temperature to drop along curve CD. On examining the phase rule equation for a condensed system, it is seen that this one-component system has one degree of freedom (univariant) in regions AB and CD. There is only one phase present – either liquid in the region AB or solid in the region CD:

$$F = C - P + 1 = 1 - 1 + 1 = 1$$

Therefore the temperature can be varied independently without the disappearance or appearance of a phase. In region BC, however, there are two phases present – liquid and solid – and the number of the degrees of freedom will be:

$$F = C - P + 1 = 1 - 2 + 1 = 0$$

Therefore the system is nonvariant. Consequently the temperature must be constant as long as two phases are present, but when all the liquid has solidified to a solid only one phase will be present and the system will again become univariant.

Curve shows a cooling curve for a binary system consisting of two metals forming a solid solution. Section AB of the curve is similar in character to that, but during the freezing period the temperature does not remain constant but drops along line BC until the whole mass as completely solidified. The application of the phase rule for section BC shows that the system is univariant; therefore the temperature will vary independently.

$F = C - P + 1 + 2 - 2 + 1 = 1$

The slope of the cooling curve, however, will change due to the evolution of latent heat of crystallization. From point C on there will be only one solid phase and the temperature falls along line CD. The melting or freezing range of alloys is due to the changes in the composition of the solid and liquid phase and naturally results in variable freezing or melting points.

A cooling curve of a binary system whose two components are completely soluble in the liquid state but entirely insoluble in the solid state. The liquid cools along line AB until temperature B is reached where a component that is in excess will crystallize and the temperature will fall along line BC, having a slope different than that of line AB.

At point C the liquid composition has been reached at which the two components crystallize simultaneously from the solution and the temperature remains constant until all the liquid solidifies. This is known as the eutectic reaction. Figure.

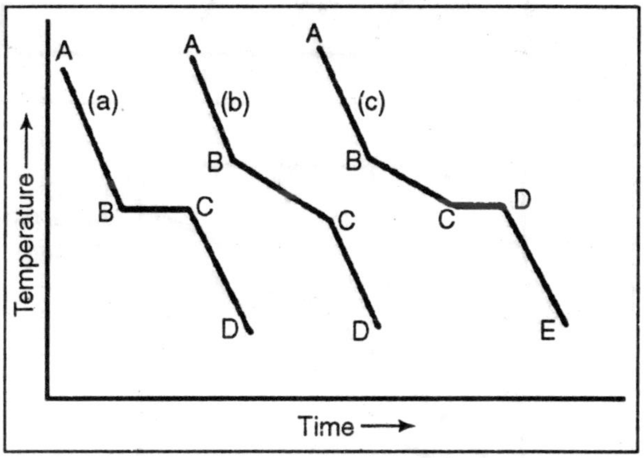

Fig. Cooling Curves: (a) Pure Compound; (b) Binary Solid Solution; (c) Binary eutectic system.

In addition to these types of cooling curves further modifications can make possible somewhat differently shaped curves.

SOLID SOLUTION EQUILIBRIUM DIAGRAMS

A plot of an equilibrium diagram for a solid solution can be made from a series of cooling curves obtained for different alloy compositions, as illustrated by the copper-nickel alloys. The temperatures corresponding to the upper points on the cooling curves (B, B1, B2,...), when plotted against suitable alloy compositions, give a curve called the "liquidus line". A plot of temperatures corresponding to lower points (C, C1, C2,...) against alloy compositions gives a curve called the "solidus line".

The solidus line represents the melting points of the different solid solutions whereas the liquid line is the freezing point curve. Points B and B4 correspond to the melting points of pure components. The region between the liquidus and solidus is a two-phase region in which solid crystals of homogeneous solid solution are in equilibrium with a liquid of suitable composition. Above the liquidus line there will be one liquid phase whereas below the solidus line there will be only one solid phase.

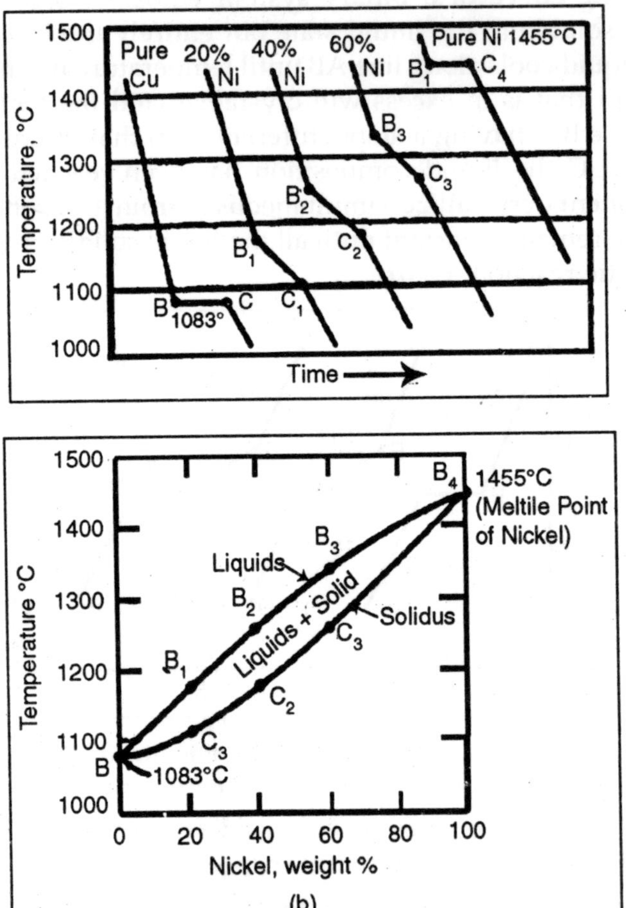

Fig. Plotting Equilibrium Diagrams from Cooling Curves for Cu-Ni Solid Solution alloys. (a) Cooling curves; (b) equilibrium diagram.

In figure we find that addition of Cu to Ni decreases its melting point. This is a commonly observed phenomenon. For example, we can depress the freezing point of water by adding common salt. At the same time, however, we find addition of Ni to Cu increases the melting point of Cu. This seems unusual and an explanation is in order.

We must first recognize that vapour pressures of a solid and a coexisting liquid are equal at the melting point. The explanation of change of melting point lies in the vapour pressures of the liquid and solid phases. If solute atoms are soluble in the liquid but insoluble in the solid, it is clear that in the liquid the attraction of solvent for solute atoms must be greater than that of solvent atoms for themselves – otherwise the two kinds of atoms would not mix.

This greater attraction means that fewer solvent atoms escape per second at the surface of the liquid – thus the vapour pressure of the liquid phase is lowered, but the vapour pressure of the solid phase is unchanged. The melting point may then either rise or fall depending on the relative degree of lowering in each phase. It should be emphasized that the above reasoning and results hold equally well for other phase changes, including polymorphic changes, as for melting and freezing.

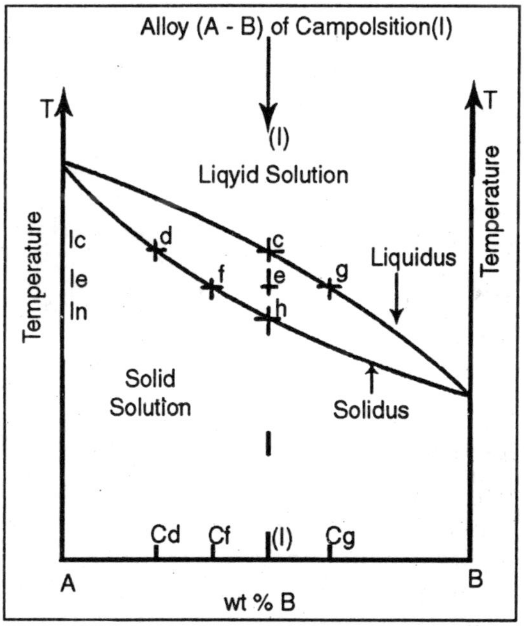

Fig. Behaviour of an Alloy of Composition.

The complete phase diagram makes it possible to predict the state of any alloy in the system at any temperature included in the diagram, if sufficient time is allowed for the system to reach equilibrium. In normal operation, however, it is fairly rare that a system reaches (or sometimes even approaches) true equilibrium, especially at lower temperatures. The equilibrium diagram is, even then, still of considerable importance since it can serve as a guide in predicting the behaviour of the system under nonequilibrium conditions. Consider the behaviour of an alloy of composition with the elements A and B, initially completely liquid, which is cooled to room temperature. As temperature decreases, there is no discernible change until the liquidus is reached (i.e., Tc). At any lower temperature, solid must appear. The composition of the solid is given by the solidus line (point d) at that temperature. Thus the first solid is much richer in A than the liquid, which is still essentially the composition of alloy I, although it is slightly richer in B than before.

The liquid now has a slightly lower melting point. As more heat is removed additional solid is formed with a composition following the solidus, whereas the liquid becomes increasingly richer in B with composition following the liquidus. A decrease in temperature to Te causes the melt to precipitate solid of f composition which both encases the existing crystals and forms new, separate crystals. For equilibrium at this temperature diffusion must take place between the d cores and the f encasements. In addition, for the entire solid to be of f composition, some B atoms from the liquid must diffuse into the A-rich centre of the crystals since diffusion between d and f compositions can only result in a composition intermediate to d and f, not in composition f. A schematic representation of these solid and liquid combinations is given in figure.

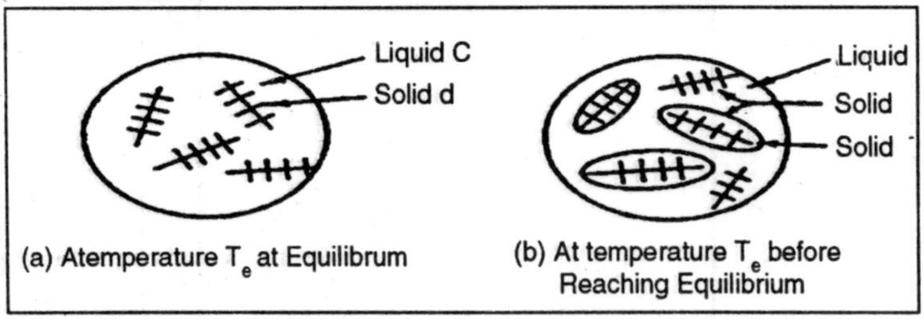

Fig. Schematic Representation of the Physical Condition of Alloy I Sketched.

If diffusion keeps pace with crystal growth (i.e., if true equilibrium is constantly maintained), melt composition moves downward along the liquidus and solid composition moves downward along the solidus. Under the microscope the separating solid appears the same as a pure metal.

This continues until temperature Th is reached. At that temperature, under equilibrium conditions, solidification is complete and on further cooling there are no further discernible changes in the solid. If the temperature of the alloy is raised from room temperature whiled constantly maintaining equilibrium to the point where the alloy is completely molten, the behaviour is exactly the reverse of the cooling behaviour.

INTERPRETATION OF PHASE DIAGRAMS

From the above discussion we can draw two useful conclusions which are the only rules necessary for interpreting equilibrium diagrams of binary systems.

Rule 1- Phase Composition

To determine the composition of phases which are stable at a given temperature we draw a horizontal line at the given temperature. The projections (upon the abscissa) of the intersections of the isothermal line with the liquidus and the solidus give the compositions of the liquid and solid, respectively, which coexist in equilibrium at that temperature. For example, draw a horizontal temperature line through temperature Te in figure. The Te line intersects the solidus at f and the liquidus at g, indicating solid composition of f% of B and (100–f)% of A. The liquid composition at this temperature is g% of B and (100–g)% of A. This line in a two-phase region is known as a tie line because it connects or "ties" together lines of one-fold saturation – i.e., the solid is saturated with respect to B and the liquid is saturated with respect to A.

Rule 2 - The Lever Rule

To determine the relative amounts of the two phases, erect an ordinate at a point on the composition scale which gives the total or overall composition of the alloy. The intersection of this composition vertical and a given isothermal line is the fulcrum of a simple lever system. The relative lengths of the lever arms multiplied by the amounts of the phase present must balance. As an illustration, consider alloy in figure. The composition vertical is erected at alloy with a composition of e% of B and (100–e)% of A. This composition vertical intersects the temperature horizontal (Te) at point e. The length of the line "f–e–g" indicates the total amount of the two phases present. The length of line "e–g" indicates the amount of solid. In other words:

$$\frac{eg}{fg} \times 100 = \%\,\text{of solid Present}$$

$$\frac{fg}{fg} \times 100 = \%\text{of liquid Present}$$

These two rules give both the composition and the relative quantity of each phase present in a two-phase region in any binary system in equilibrium regardless of physical form of the two phases. The two rules apply only to two-phase regions.

ISOMORPHOUS SYSTEMS

An isomorphous system is one in which there is complete intersolubility between the two components in the vapour, liquid and solid phases, as shown in figure. The Cu-Ni system is both a classical and a practical example since the monels, which enjoy extensive commercial use, are Cu-Ni alloys. Many practical materials systems are isomorphous.

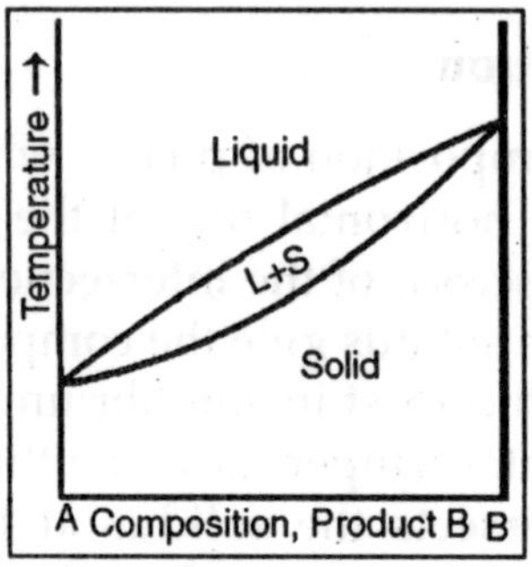

Fig. Schematic Phase Diagram for a Binary System, A-B, Showing Complete Intersolubility (Isomorphism) In all Phases.

INCOMPLETE SOLUBILITY

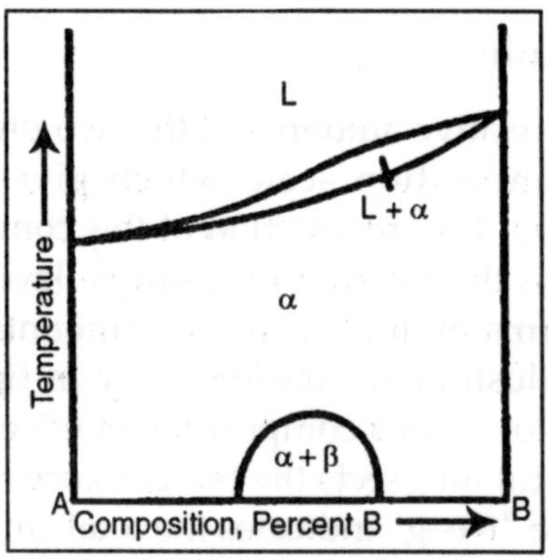

Fig. Phase Diagram of A-B System Showing a Small Solubility Gap.

In many systems there is a tendency toward incomplete intersolubility below the transformation temperature and we find a solubility gap (e.g., Cu-Pd system) similar to that in figure. If there is a greater tendency toward incomplete intersolubility we may find a larger gap, as in figure. Diagrams similar to this are found in the Au-Cu and Au-Ni systems and others. If the tendency for formation of a solubility gap increases beyond that in figure, we might find the singular, but possible, case in which the maximum of the solubility gap and the minima of the liquidus and solidus have a common horizontal tangent. At the point of tangency, three phases (α, β and liquid) would be in equilibrium without violating the phase rule.

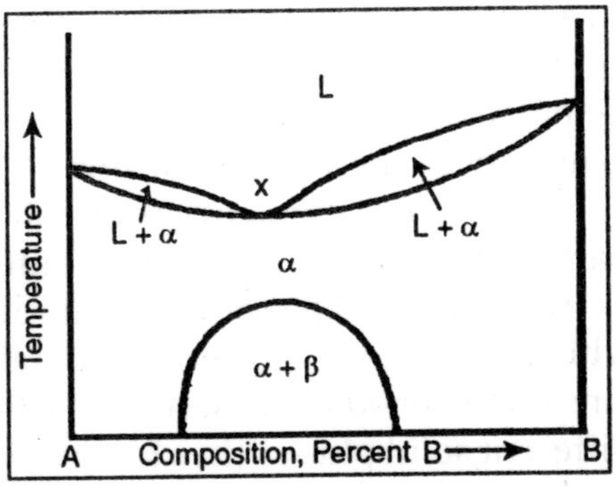

Fig. Phase Diagram of A-B System Showing a Larger Gap

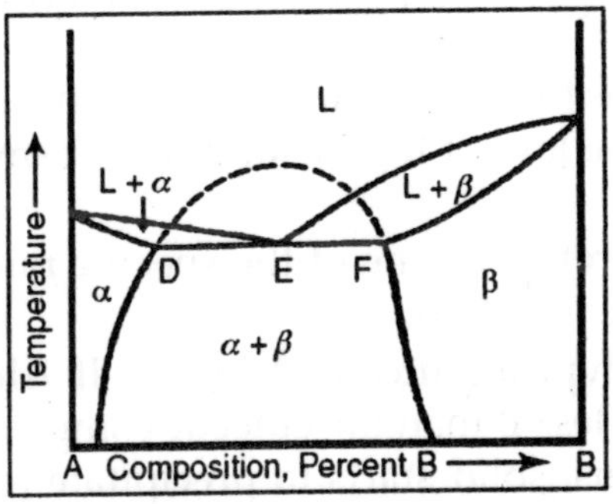

Fig. Phase Diagram of A-B System Showing a Large Solubility Gap. This is the Origin of the Eutectic-type Phase Diagram.

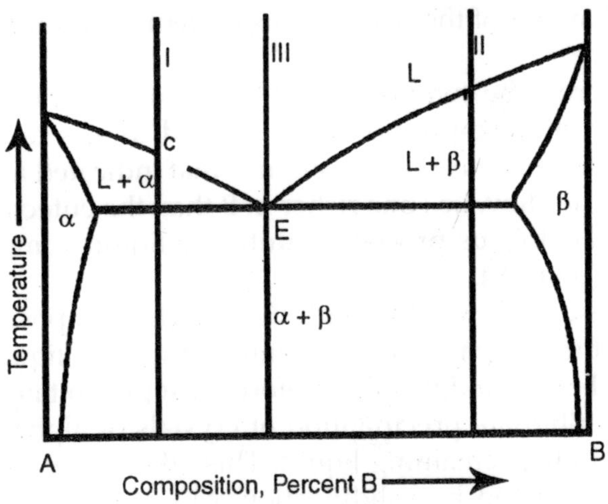

Fig. Phase Diagram of A-B System Showing a Typical Eutectic Diagram.

EUTECTIC SYSTEMS

If intersolubility decreases further, implying an increased solubility gap, then the upper transformation curves effectively overlap the gap and we obtain a diagram of the kind shown in figure. This is a schematic representation of the eutectic type of diagram.

All lines in figure, except the horizontal line D–E–F, are one-phase region boundaries (i.e., transformation curves or solubility limit curves). The line D–E–F is part of the solidus. The work "eutectic" (coming from the Greek, meaning "most fusible") has at least four different, but related, meanings – a fact which sometimes causes considerable confusion.

- Eutectic means the reversible, isothermal reaction of a liquid which forms two different solid phases (in a binary system) upon cooling – i.e., L → a + b.
- Eutectic means the composition of the alloy which undergoes this reaction.
- Eutectic means the temperature at which the eutectic reaction takes place – i.e., the lowest liquid/solid transformation temperature in the system.
- Eutectic also refers to one of the specific microstructures formed by the eutectic reaction.

Figure shows a typical eutectic diagram. If Alloy I (a hypoeutectic alloy, i.e., one having less B than the eutectic composition) is cooled under equilibrium conditions, solid starts to precipitate (or separate) from the liquid when the temperature is that of the intersection of the composition vertical with the liquidus line (point c). The solid phase is not pure A, but is a primary solid solution of B in A, exactly as discussed for Alloy I Figure. The behaviour of the two systems is identical until Alloy I is cooled to Te.

In other words, the composition of the solid moves downward along the solidus and the composition of the liquid moves downward along the liquidus until the liquid composition is that indicated at point E. Alloy II (a hypereutectic alloy, i.e., one richer in B than the eutectic composition) will follow an analogous procedure until the liquid composition is that of the eutectic (point E).

In either case, or if we start with the eutectic composition, we arrive at point E on the diagram. At this temperature the liquid is saturated with respect to both A and B and therefore must precipitate αand βphases. This occurs by alternate precipitation of crystals of alpha and of beta at various points in the remaining liquid. This takes place as rapidly as the heat of fusion, released by solidification, can be removed. The process continues until all the liquid has solidified. The temperature and composition (but not the relative quantities) of all three phases remain constant during solidification.

Representative microscopic structures for these three alloys are shown in figure. Alloy II has primary crystals of β surrounded by eutectic. Alloy II has primary crystals of βsurrounded by eutectic. Very crudely speaking, these two structures resemble pebbles in concrete. The eutectic structure itself, whether in Alloy III or in either of the above allb alloys, is an intimate mechanical mixture of α and β.

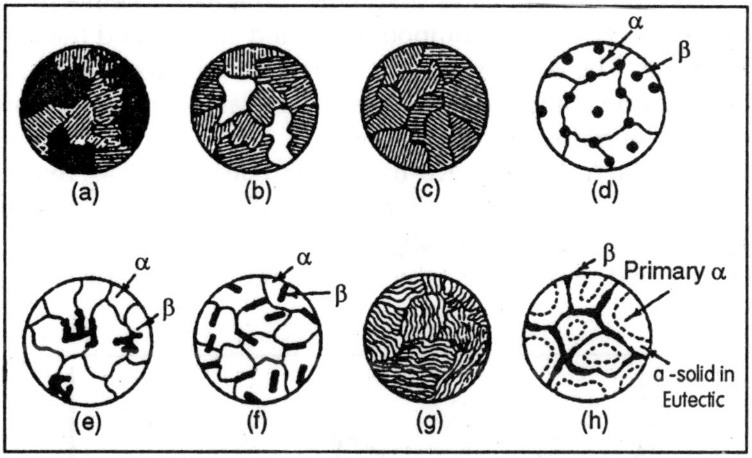

Fig. Schematic Representation Possible in Eutectic Structures. (a), (b) and (c) are Alloys; (d) Nodular; (e) Chinese Script; (f) Acicular; (g) Lamellar; and (h) Divorced.

The lamellar, or alternate platelet, type of eutectic structure is most commonly found. One of three additional structures is often found when the eutectic composition is close to the composition of one of the solid solutions composing the eutectic.

These structures are the "Chinese script", acicular (needle-like) and nodular (spheroidal) structures. A "divorced eutectic" is sometimes also found. In this case there is no distinct boundary between the primary solid solution and the portion of the same solid solution in the eutectic. In other words, there is a tendency for the two phases to separate into large crystals so that there is little or no resemblance to one of the normal types of eutectic structure.

EQUILIBRIUM DIAGRAMS WITH INTERMEDIATE COMPOUNDS

Two components may also form an intermediate compound at certain composition. The melting point of such an intermediate compound or compounds may be above, below or between the melting points of the pure components. These compounds have a different crystalline structure from that of the parent substances and consequently form another solid phase in the system.

Figure illustrates such a system in which the intermediate compound is formed from two components, A and B. Such a binary compound AB forms reversibly as the result of phase reactions between two components and exhibits a composition constant and independent of pressure and temperature.

The diagram also indicates that there are now two eutectic points, E and E′, formed between the compound AB and solid A and the compound AB and solid B. The compound has a congruent melting point, giving a liquid of the same composition as the crystal, and it does not form solid solutions with either component A or B.

The systems A-AB and AB-B may be considered as two separate systems within the boundary system A-B, and they will behave similarly, as exemplified by copper-zinc alloys, in which the maximum solubility has been reached and above which the components form intermetallic compounds.

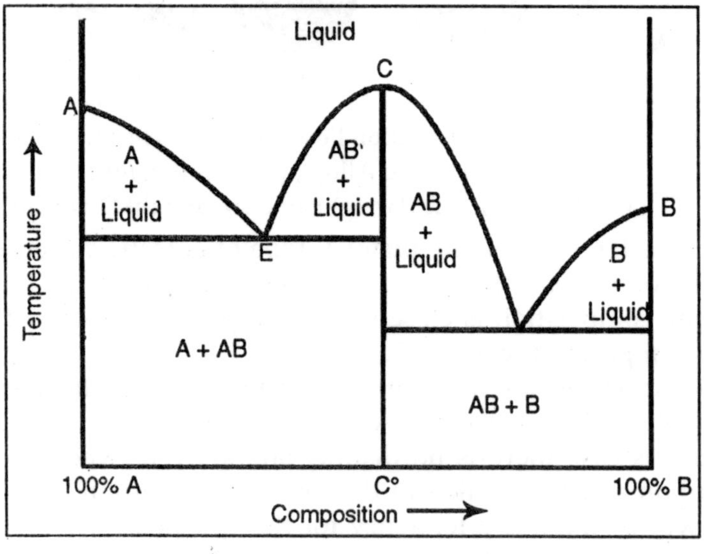

Fig. Binary System Showing an Intermediate Compound. C is the Melting Point (Maximum) of the Compound AB having the Composition C′. E is the Eutectic of Solid A and Solid AB. E′ is the Eutectic of Solid AB and Solid B.

SYNTHESIS OF SOLID STATE CHEMISTRY

The construction of organic–inorganic nano structure materials is an important target of modern materials research. The purpose behind such research is to develop functional materials. The smectite group, which includes montmorillonite, has enjoyed wide attention due to its interesting adsorptive/intercalation and cation-exchange properties. Organoclay has received widespread attention in a number of industrial sectors including waste treatment catalysis, and electrodemodifiers. The synthesis, characterization and properties of smectite–organic intercalation compounds have been reported extensively.

The process often used for the preparation of organoclay is an intercalation into the clay in suspension state. These reactions are characterized by long time treatment and excessive use of organic solvent. Therefore, it is interesting to develop and test another efficient process of preparation of organoclay. A process, which is not very frequently used and which allows the possibility of preparing compounds which are not accessible from solutions, is the socalled solid–solid reaction.

The latter is a mechanochemical adsorption that occurs between powders in the solid state. Solid-state intercalation of organic guest species into layered vanadium phosphorus oxide, zeolites and layered clay minerals have been reported as a way of utilizing host structures. Solidstate intercalation can be split into two stages: one is the diffusion of guest species from the outer surfaces of guest solids and the other is the penetration of the guest species into the interlayer spaces. Intercalation of pyridine and related derivatives in clays has been widely studied for several decades.

Their intercalation can be achieved from the neutral molecules by interaction of the pyridine ring with the interlayer cations through water bridges and it can also be achieved by cation exchange from the corresponding protonated (pyridinium) derivatives. What is interesting is the fact that intercalated pyridines can adopt different orientations in the interlayer space depending on both the structure of such molecules and the nature of the interlayer cation. We report here the intercalation of 2-mercaptopyridine into Na-, Co(II)- and Al(III)-montmorillonite by solid–solid reaction. We have chosen this organic molecule for two principal reasons:

(i) It is an interesting derivative of pyridine which contains functional groups (thiolate-S and aromatic-N) and therefore, it may interact with clay in different orientations and configurations,

(ii) It has the ability to both chelate and bridge transition metals. We have characterized the intercalated products by Xray diffraction (XRD), atomic force microscopy (AFM), infrared spectroscopy (FTIR), azote adsorption–desorption and thermo XRD.

Chapter 11

Ionizing Radiation

IONIZING VERSUS NON-IONIZING RADIATION

We live in a sea of radiation. In recent years, people have learned to fear the effects of radiation. They don't want to live near nuclear reactors. They are frightened by reports of links between excess exposure to sunlight and skin cancer. They are afraid of the leakage from microwave ovens, or the radiation produced by their television sets.

Several factors combine to heighten the public's anxiety about both the short-range and long-range effects of radiation. Perhaps the most important source of fear is the fact that radiation can't be detected by the average person. Furthermore, the effects of exposure to radiation might not appear for months or years or even decades.

To understand the biological effects of radiation we must first understand the difference between ionizing radiation and non-ionizing radiation. In general, two things can happen when radiation is absorbed by matter: excitation or ionization.

- Excitation occurs when the radiation excites the motion of the atoms or molecules, or excites an electron from an occupied orbital into an empty, higher-energy orbital.
- Ionization occurs when the radiation carries enough energy to remove an electron from an atom or molecule.

Because living tissue is 70-90% water by weight, the dividing line between radiation that excites electrons and radiation that forms ions is often assumed to be equal to the ionization of water: 1216 kJ/mol. Radiation that carries less energy can only excite the water molecule. It is therefore called non-ionizing radiation. Radiation that carries more energy than 1216 kJ/mol can remove an electron from a water molecule, and is therefore called ionizing radiation. The table below contains estimates of the energies of various kinds of radiation. Radio waves, microwaves, infrared radiation, and visible light are all forms of non-ionizing radiation. X-rays, γ-rays, and α- and β -particles are forms of ionizing radiation. The dividing line between ionizing and non-ionizing radiation in the electromagnetic spectrum falls in the ultraviolet portion of the spectrum. It is therefore useful to divide the UV spectrum into two categories: UV_A and UV_B. Radiation at the high-energy end of the UV spectrum can be as dangerous as x-rays or γ-rays.

Table. Energies of Ionizing and Non-Ionizing Forms of Radiation.

Radiation	Typical Frequency (s^{-1})	Typical Energy (kJ/mol)		
Particles				
α-particles		4.1×10^8		
β -particles		1.5×10^7		
Electromagnetic radiation				ionizing
cosmic rays	$6 \times 10^{21} s^{-1}$	2.4×10^9	}	radiation
γ -rays	$3 \times 10^{20} s^{-1}$	1.2×10^8		
x-rays	$3 \times 10^{17} s^{-1}$	1.2×10^5		
ultraviolet	$3 \times 10^{15} s^{-1}$	1200		
visible	$5 \times 10^{14} s^{-1}$	200		
infrared	$3 \times 10^{13} s^{-1}$	12		non-ionizing
microwaves	$3 \times 10^9 s^{-1}$	1.2×10^{-3}	}	radiation
radio waves	$3 \times 10^7 s^{-1}$	1.2×10^{-5}		

When ionizing radiation passes through living tissue, electrons are removed from neutral water molecules to produce H_2O^+ ions. Between three and four water molecules are ionized for every 1.6×10^{-17} joules of energy absorbed in the form of ionizing radiation.

$$H_2O \rightarrow H_2O^+ + e^-$$

The H_2O^+ ion should not be confused with the H_3O^+ ion produced when acids dissolve in water. The H_2O^+ ion is an example of a free radical, which contains an unpaired valence-shell electron. Free radicals are extremely reactive. The radicals formed when ionizing radiation passes through water are among the strongest oxidizing agents that can exist in aqueous solution.

At the molecular level, these oxidizing agents destroy biologically active molecules by either removing electrons or removing hydrogen atoms. This often leads to damage to the membrane, nucleus, chromosomes, or mitochondria of the cell that either inhibits cell division, results in cell death, or produces a malignant cell.

BIOLOGICAL EFFECTS OF IONIZING RADIATION

From the time that radioactivity was discovered, it was obvious that it caused damage. As early as 1901, Pierre Curie discovered that a sample of radium placed on his skin produced wounds that were very slow to heal. What some find surprising is the magnitude of the difference between the biological effects of non-ionizing radiation, such as light and microwaves, and ionizing radiation, such as high-energy ultraviolet radiation, x-rays, γ-rays, and α- or β -particles.

Radiation at the low-energy end of the electromagnetic spectrum, such as radio waves and microwaves, excites the movement of atoms and molecules, which is equivalent to heating the sample. Radiation in or near the visible portion of the spectrum excites electrons into higher-energy orbitals. When the electron eventually falls back to a lower-energy state, the excess energy is given off to neighboring molecules in the form of heat. The principal effect of non-ionizing radiation is therefore an increase in the temperature of the system.

We experience the fact that biological systems are sensitive to heat each time we cook with a microwave oven, or spend too long in the sun. But it takes a great deal of non-ionizing radiation to reach dangerous levels.

We can assume, for example, that absorption of enough radiation to produce an increase of about 6C in body temperature would be fatal. Since the average 70-kilogram human is 80% water by weight, we can use the heat capacity of water to calculate that it would take about 1.5 million joules of non-ionizing radiation to kill the average human. If this energy was carried by visible light with a frequency of 5×10^{14} s^{-1}, it would correspond to absorption of about seven moles of photons.

Ionizing radiation is much more dangerous. A dose of only 300 joules of x-ray or γ-ray radiation is fatal for the average human, even though this radiation raises the temperature of the body by only 0.001C. α-particle radiation is even more dangerous; a dose equivalent to only 15 joules is fatal for the average human. Whereas it takes seven moles of photons of visible light to produce a fatal dose of non-ionizing radiation, absorption of only 7×10^{-10} moles of the α-particles emitted by ^{238}U is fatal.

There are three ways of measuring ionizing radiation.

- Measure the activity of the source in units of disintegrations per second or curies, which is the easiest measurement to make.
- Measure the radiation to which an object is exposed in units of roentgens by measuring the amount of ionization produced when this radiation passes through a sample of air.
- Measure the radiation absorbed by the object in units of radiation absorbed doses or "rads." This is the most useful quantity, but it is the hardest to obtain.

One radiation absorbed dose, or rad, corresponds to the absorption of 10^{-5} joules of energy per gram of body weight. Because this is equivalent to 0.01 J/kg, one rad produces an increase in body temperature of about 2×10^{-6}C.

At first glance, the rad may seem to be a negligibly small unit of measurement. The destructive power of the radicals produced when water is ionized is so large, however, that cells are inactivated at a dose of 100 rads, and a dose of 400 to 450 rads is fatal for the average human.

Not all forms of radiation have the same efficiency for damaging biological organisms. The faster energy is lost as the radiation passes through the tissue, the more damage it does. To correct for the differences in radiation biological effectiveness (RBE) among various forms of radiation, a second unit of absorbed dose has been defined. The roentgen equivalent man, or rem, is the absorbed dose in rads times the biological effectiveness of the radiation.

rems = rads x RBE

Values for the RBE of different forms of radiation are given in the table below.

The Radiation Biological Effectiveness of Various Forms of Radiation

Radiation	RBE
x-rays and γ-rays	1
b - particles with energies larger than 0.03 MeV	1
b - particles with energies less than 0.3 MeV	1.7
thermal (slow-moving) neutrons	3
fast-moving neutrons or protons	10
α-particles or heavy ions	20

Estimates of the per capita exposure to radiation in the United States are summarized in the table below. These estimates include both external and internal sources of natural background radiation.

Average Whole-Body Exposure Levels for Sources of Ionizing Radiation

Source		Per Capita Dose (rems / y)
Natural background		0.082
Medical x-rays		0.077
Nuclear test fallout		0.005
Consumer and industrial products		0.005
Nuclear power industry		0.001
	Total:	0.170

External sources include cosmic rays from the sun and α-particles or α-rays emitted from rocks and soil. Internal sources include nuclides that enter the body when we breathe (^{14}C, ^{85}Kr, ^{220}Rn, and ^{222}Rn) and through the food chain (^{3}H, ^{14}C, ^{40}K, ^{90}Sr, ^{131}I, and ^{137}Cs). The actual dose from natural radiation depends on where one lives.

People who live in the Rocky Mountains, for example, receive twice as much background radiation as the national average because there is less atmosphere to filter out the cosmic rays from the sun. The average dose from medical x-rays has decreased in recent years because of advances in the sensitivity of the photographic film used for x-rays. Radiation from nuclear test fallout has also decreased as a result of the atmospheric nuclear test ban.

The threat of fallout from the testing of nuclear weapons can be appreciated by noting that a Chinese atmospheric test in 1976 led to the contamination of milk in the Harrisburg, PA, vicinity at a level of 300 pCi (3.00×10^{-10} Ci) per lite. This was about eight times the level of contamination (41 pCi per liter) that resulted from the accident at Three Mile Island.

The contribution to the radiation absorbed dose from consumer and industrial products includes radiation from construction materials, x-rays emitted by television sets, and inhaled tobacco smoke. The most recent estimate of the total radiation emitted from the mining and milling of uranium, the fabrication of reactor fuels, the storage of radioactive wastes, and the operation of nuclear reactors is less than 0.001 rem per year.

The total dose from ionizing radiation for the average American is about 0.170 rem per year. The Committee on the Biological Effects of Ionizing Radiation of the National Academy of Sciences recently estimated that an increase in this dose to a level of 1 rem per year would result in 169 additional deaths from cancer per million people exposed. This can be compared with the 170,000 cancer deaths that would normally occur in a population this size that was not exposed to this level of radiation.

The principal effect of low doses of ionizing radiation is to induce cancers, which may take up to 20 years to develop. What is the effect of high doses of ionizing radiation? Cells that are actively dividing are more sensitive to radiation than cells that aren't. Thus, cells in the liver, kidney, muscle, brain, and bone are more resistant to radiation than the cells of bone marrow, the reproductive organs, the epithelium of the intestine, and the skin, which suffer the most damage from radiation.

Damage to the bone marrow is the main cause of death at moderately high levels of exposure (200 to 1000 rads). Damage to the gastrointestinal tract is the major cause of death for exposures on the order of 100 to 10,000 rads. Massive damage to the central nervous system is the cause of death from extremely high exposures (over 10,000 rads).

NATURAL VERSUS INDUCED RADIOACTIVITY

NATURAL RADIOACTIVITY

The vast majority of the nuclides found in nature are stable. If our planet is 4.6 billion years old, the only radioactive isotopes that should remain are members of three classes.

- Isotopes with half-lives of at least 10^9 years, such as ^{238}U.
- Daughter nuclides produced when long-lived radioactive nuclides decay, such as the ^{234}Th ($t_{1/2}$ = 24.1 days) produced by the α-decay of ^{238}U.
- Nuclides such as ^{14}C that are still being synthesized.

One factor that influences the stability of a nuclide is the ratio of neutrons to protons. (Nuclei that contain either too many or too few neutrons are unstable.) Another factor that affects the stability of nuclides can be understood by examining patterns in the numbers of protons and neutrons in stable nuclides.

Half of the elements in the periodic table must have an odd number of protons because atomic numbers that are odd are just as likely to occur as those that are even. In spite of this, about 80% of the stable nuclides have an even number of protons. Very few elements with an odd atomic number have more than one stable isotope.

Stable isotopes abound, however, among elements with even atomic numbers. Ten stable isotopes are known for tin ($Z = 50$), for example. It is also interesting to note that 91% of the stable isotopes of elements with an odd number of protons have an even number of neutrons. These observations suggest that certain combinations of protons and neutrons are particularly stable.

There are magic numbers of electrons. Electron configurations with 2, 10, 18, 36, 54, and 86 electrons are unusually stable. There also seem to be magic numbers of neutrons and protons. Nuclei with 2, 8, 20, 28, 50, 82, or 126 protons or neutrons are unusually stable. This observation explains the anomalously large binding energies observed for ^{4}He, ^{16}O, and ^{20}Ne. In each case, the nuclide has an even number of both protons and neutrons. ^{20}Ne has a magic number of nucleons when both protons and neutrons are counted. ^{4}He and ^{16}O have magic numbers of both protons and neutrons. The resulting stability of the ^{4}He nucleus might explain why so many heavy nuclei undergo α-particle decay by ejecting an $^{4}He^{2+}$ ion from the nucleus of the atom.

If nuclei tend to be more stable when they have even numbers of protons and neutrons, it isn't surprising that nuclides with an odd number of both protons and neutrons are unstable. ^{40}K is one of only five naturally occurring nuclides that contain both an odd number of protons and an odd number of neutrons. This nuclide simultaneously undergoes the electron capture and positron emission expected for neutron-poor nuclides and the electron emission observed with neutron-rich nuclides.

Only 18 radioactive isotopes with atomic numbers of 80 or less can be found in nature. With the exception of ^{14}C, which is continuously synthesized in the atmosphere, all these elements have lifetimes longer than 10^9 years. Although these isotopes all undergo radioactive decay, they decay so slowly that reasonable quantities are still present, 4.6 billion years after the planet was formed.

Another 45 natural radioactive isotopes have atomic numbers larger than 80. These nuclides fall into three families. The parent nuclide is ^{232}Th, which undergoes α-decay to form ^{228}Ra. The product of this reaction decays by β -emission to form ^{228}Ac, which decays to ^{228}Th, and so on, until the stable ^{208}Pb isotope is formed. This family of radionuclides is called the 4n series, because all its members have a mass number that can be divided by 4.

A second family of radioactive nuclei starts with ^{238}U and decays to form the stable ^{206}Pb isotope. Every member of this series has a mass number that fits the equation 4n + 2. The third family, known as the 4n + 3 series, starts with ^{235}U and decays to ^{207}Pb. A 4n + 1 series once existed, which started with ^{237}Np and decayed to form the only stable isotope of bismuth, ^{209}Bi. The half-life of every member of this series is less than 2 x 10^6 years, however, so none of the nuclides produced by the decay of neptunium remain in detectable quantities on the earth.

INDUCED RADIOACTIVITY

In 1934, Irene Curie, the daughter of Pierre and Marie Curie, and her husband, Frederic Joliot, announced the first synthesis of an artificial radioactive isotope. They bombarded a thin piece of aluminum foil with α-particles produced by the decay of polonium and found that the aluminum target became radioactive. Chemical analysis showed that the product of this reaction was an isotope of phosphorus.

$$ {}^{27}_{13}\text{AI} + {}^{4}_{2}\text{He} \rightarrow {}^{30}_{15}\text{P} + {}^{1}_{0}\text{n} $$

In the next 50 years, more than 2000 other artificial radionuclides were synthesized.

A shorthand notation has been developed for nuclear reactions such as the reaction discovered by Curie and Joliot. The parent (or target) nuclide and the daughter nuclide are separated by parentheses that contain the symbols for the particle that hits the target and the particle or particles released in this reaction.

$$ {}^{27}_{13}\text{AI}(\alpha,\text{n}){}^{30}_{15}\text{P} $$

The nuclear reactions used to synthesize artificial radionuclides are characterized by enormous activation energies. Three devices are used to overcome these activation energies: linear accelerators, cyclotrons, and nuclear reactors.

Linear accelerators or cyclotrons can be used to excite charged particles such as protons, electrons, α-particles, or even heavier ions, which are then focused on a stationary target. The following reaction, for example, can be induced by a cyclotron or linear accelerator.

$$ {}^{24}_{12}\text{Mg} + {}^{2}_{1}\text{H} \rightarrow {}^{22}_{11}\text{Na} + {}^{4}_{2}\text{He} $$

Because these reactions involve the capture of a positively charged particles, they usually produce a neutron-poor nuclide.

Artificial radionuclides are also synthesized in nuclear reactors, which are excellent sources of slow-moving, or thermal neutrons. The absorption of a neutron usually results in a neutron-rich nuclide. The following neutron absorption reaction occurs in the cooling systems of nuclear reactors cooled with liquid sodium metal.

$$ {}^{23}_{11}\text{Na} + {}^{1}_{0}\text{n} \rightarrow {}^{24}_{11}\text{Na} + \gamma $$

In 1940, absorption of thermal neutrons was used to synthesize the first elements with atomic numbers larger than the heaviest naturally occurring element, uranium. The first of these truly artificial elements were neptenium and plutonium, which were synthesized by McMillan and Abelson by irradiating ^{238}U with neutrons to form ^{239}U

$$^{238}_{92}U + ^{1}_{0}n \rightarrow ^{239}_{92}U + \gamma(\text{Neutron Capture})$$

which undergoes β^--decay to form ^{239}Np and then ^{239}Pu.

$$^{238}_{92}U \rightarrow ^{239}_{92}Np + ^{0}_{-1}e$$

$$^{238}_{92}NP \rightarrow ^{239}_{94}Pu + ^{0}_{-1}e$$

Larger bombarding particles were eventually used to produce even heavier transuranium elements.

$$^{253}_{99}Es + ^{4}_{2}He \rightarrow ^{256}_{101}Md + ^{1}_{0}n$$

$$^{246}_{96}Cm + ^{12}_{6}C \rightarrow ^{254}_{102}No + 4\,^{1}_{0}n$$

The half-lives for α-decay and spontaneous fission decrease as the atomic number of the element increases. Element 104, for example, has a half-life for spontaneous fission of 0.3 seconds. Elements therefore become harder to characterize as the atomic number increases. Recent theoretical work has predicted that a magic number of protons might exist at Z = 114.

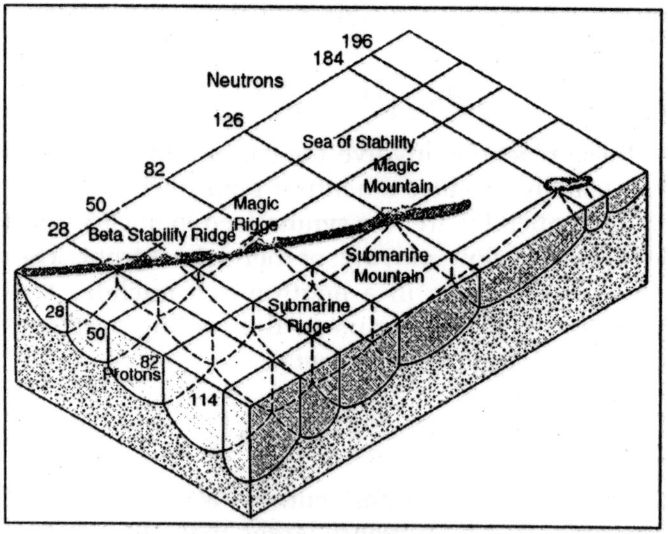

Fig. Spontaneous Fission.

This work suggests that there is an island of stability in the sea of unstable nuclides, as shown in the figure below. If this theory is correct, superheavy elements could be formed if we could find a way to cross the gap between elements Z = 109 through Z = 114. There is some debate about the number of neutrons needed to overcome the proton-proton repulsion in a nucleus with 114 protons. The best estimates suggest that at least 184 neutrons, and perhaps as many as 196, would be needed. It is not an easy task to bring together two particles that give both the correct number of total protons and the necessary neutrons to produce a nuclide with a half-life long enough to be detected.

If we start with a relatively long-lived parent nuclide, such as ^{251}Cf ($t_{1/2}$ = 800 y) and bombard this nucleus with a heavy ion, such as ^{32}S, we can envision producing a daughter nuclide with the correct atomic number, but the mass number would be too small by at least 16 amu.

$$^{251}_{98}Cf + ^{32}_{16}S \rightarrow ^{282}_{114}X + ^{1}_{0}n$$

An expanded periodic table for elements up to Z = 168 is shown in the figure below. Elements 104 through 112 are transition metals that fill the 6d orbitals. Elements 113 through 120 are main-group elements in which the 7p and 8s orbitals are filled.

The next subshell is the 5g atomic orbital, which can hold up to 18 electrons. There is reason to believe that the 5g and 6f orbitals will be filled at the same time. The next 32 elements are therefore grouped into a so-called superactinide series.

NUCLEAR FISSION AND NUCLEAR FUSION

NUCLEAR FISSION

The graph of binding energy per nucleon suggests that nuclides with a mass larger than about 130 amu should spontaneously split apart to form lighter, more stable, nuclides. Experimentally, we find that spontaneous fission reactions occur for only the very heaviest nuclides – those with mass numbers of 230 or more. Even when they do occur, these reactions are often very slow. The half-life for the spontaneous fission of ^{238}U, for example, is 10^{16} years, or about two million times longer than the age of our planet! We don't have to wait, however, for slow spontaneous fission reactions to occur. By irradiating samples of heavy nuclides with slow-moving thermal neutrons it is possible to induce fission reactions. When ^{235}U absorbs a thermal neutron, for example, it splits into two particles of uneven mass and releases an average of 2.5 neutrons, as shown in the figure below.

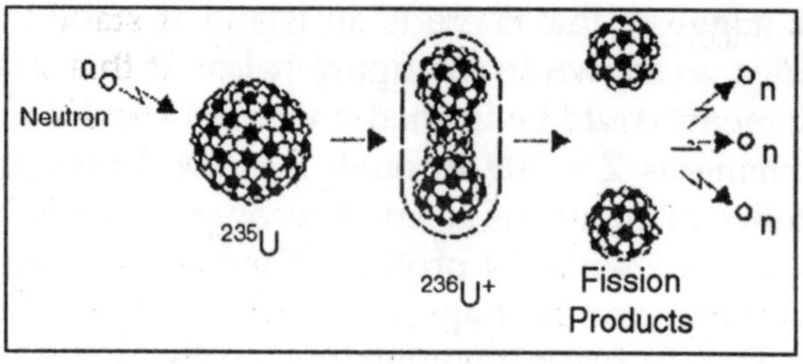

Fig. The Absorption of a Neutron by ^{238}U Induces Scillations in the Nucleus that Deform it Until it Splits into Fragments the way a Drop of Liquid Might Break into Smaller Droplets.

More than 370 daughter nuclides with atomic masses between 72 and 161 amu are formed in the thermal-neutron-induced fission of ^{235}U, including the two products shown below.

$$^{235}_{92}U + ^{1}_{0}n \rightarrow ^{139}_{56}Ba + ^{94}_{36}Kr + 3\,^{1}_{0}n$$

Several isotopes of uranium undergo induced fission. But the only naturally occurring isotope in which we can induce fission with thermal neutrons is ^{235}U, which is present at an abundance of only 0.72%. The induced fission of this isotope releases an average of 200 MeV per atom, or 80 million kilojoules per gram of ^{235}U. The attraction of nuclear fission as a source of power can be understood by comparing this value with the 50 kJ/g released when natural gas is burned.

The first artificial nuclear reactor was built by Enrico Fermi and co-workers beneath the University of Chicago's football stadium and brought on line on December 2, 1942. This reactor, which produced several kilowatts of power, consisted of a pile of graphite blocks weighing 385 tons stacked in layers around a cubical array of 40 tons of uranium metal and uranium oxide. Spontaneous fission of ^{238}U or ^{235}U in this reactor produced a very small number of neutrons.

But enough uranium was present so that one of these neutrons induced the fission of a ^{235}U nucleus, thereby releasing an average of 2.5 neutrons, which catalyzed the fission of additional ^{235}U nuclei in a chain reaction, as shown in the figure below. The amount of fissionable material necessary for the chain reaction to sustain itself is called the critical mass.

The Fermi reactor at Chicago served as a prototype for larger reactors constructed in 1943 at Oak Ridge, Tennessee, and Hanford, Washington, to produce ^{239}Pu for one of the atomic bombs dropped on Japan at the end of World War II.

As we have seen, some of the neutrons released in the chain reaction are absorbed by ^{238}U to form ^{239}U, which undergoes decay by the successive loss of two β -particles to form ^{239}Pu. ^{238}U is an example of a fertile nuclide. It doesn't undergo fission with thermal neutrons, but it can be converted to ^{239}Pu, which does undergo thermal-neutron-induced fission.

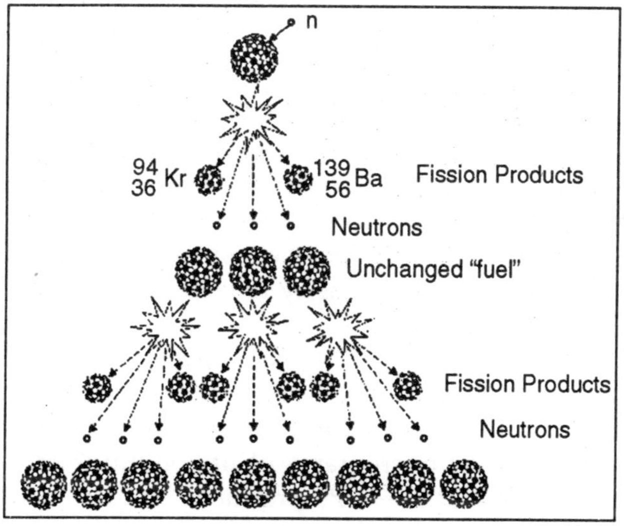

Fig. Fission Reaction.

Fission reactors can be designed to handle naturally abundant ^{235}U, as well as fuels described as slightly enriched (2–5% ^{235}U), highly enriched (20-30% ^{235}U), or fully enriched (>90% ^{235}U). Heat generated in the reactor core is transferred to a cooling agent in a closed system.

The cooling agent is then passed through a series of heat exchangers in which water is heated to steam. The steam produced in these exchangers then drives a turbine that generates electrical power. There are two ways of specifying the power of such a plant: the thermal energy produced by the reactor or the electrical energy generated by the turbines. The electrical capacity of the plant is usually about one-third of the thermal power.

It takes 10^{11} fissions per second to produce one watt of electrical power. As a result, about one gram of fuel is consumed per day per megawatt of electrical energy produced. This means that one gram of waste products is produced per megawatt per day, which includes 0.5 grams of ^{239}Pu. These waste products must be either reprocessed to generate more fuel or stored for the tens of thousands of years it takes for the level of radiation to reach a safe limit.

If we could design a reactor in which the ratio of the ^{239}Pu or ^{233}U produced to the ^{235}U consumed was greater than 1, the reactor would generate more fuel than it consumed. Such reactors are known as breeders, and commercial breeder reactors are now operating in France. The key to an efficient breeder reactor is a fuel that gives the largest possible number of neutrons released per neutron absorbed. The breeder reactors being built today use a mixture of PuO_2 and UO_2 as the fuel and fast neutrons to activate fission. Fast neutrons carry an energy of at least several KeV and therefore travel 10,000 or more times faster than thermal neutrons. ^{239}Pu in the fuel assembly absorbs one of these fast neutrons and undergoes fission with the release of three neutrons. ^{238}U in the fuel then captures one of these neutrons to produce additional ^{239}Pu.

The advantage of breeder reactors is obvious–they mean a limitless supply of fuel for nuclear reactors. There are significant disadvantages, however. Breeder reactors are more expensive to build. They are also useless without a subsidiary industry to collect the fuel, process it, and ship the ^{239}Pu to new reactors.

It is the reprocessing of ^{239}Pu that concerns most of the critics of breeder reactors. ^{239}Pu is so dangerous as a carcinogen that the nuclear industry places a limit on exposure to this material that assumes workers inhale no more than 0.2 micrograms of plutonium over their lifetimes. There is also concern that the ^{239}Pu produced by these reactors might be stolen and assembled into bombs by terrorist organizations.

The fate of breeder reactors in the United States is linked to economic considerations. Because of the costs of building these reactors and safely reprocessing the ^{239}Pu produced, the breeder reactor becomes economical only when the scarcity of uranium drives its price so high that the breeder reactor becomes cost effective by comparison. If nuclear energy is to play a dominant role in the generation of electrical energy in the 21st century, breeder reactors eventually may be essential.

Although the "pile" Fermi constructed at the University of Chicago in 1942 was the first artificial nuclear reactor, it was not the first fission reactor to exist on Earth. In 1972, a group of French scientists discovered that uranium ore from a deposit in the Oklo mine in Gabon, West Africa, contained 0.4% ^{235}U instead of the 0.72% abundance found in all other sources of this ore.

Analysis of the trace elements in the ore suggested that the amount of ^{235}U in this ore was unusually small because natural fission reactors operated in this deposit for a period of 600,000 to 800,000 years about 2 billion years ago.

NUCLEAR FUSION

The graph of binding energy per nucleon suggests another way of obtaining useful energy from nuclear reactions. Fusing two light nuclei can liberate as much energy as the fission of ^{235}U or ^{239}Pu. The fusion of four protons to form a helium nucleus, two positrons (and two neutrinos), for example, generates 24.7 MeV of energy.

$$4\,{}^{1}_{1}H \rightarrow {}^{4}_{2}He + 2\,{}^{0}_{+1}e$$

Most of the energy radiated from the surface of the sun is produced by the fusion of protons to form helium atoms within its core.

Fusion reactions have been duplicated in man-made devices. The enormous destructive power of the ^{235}U-fueled atomic bomb dropped on Hiroshima on August 6, 1945, which killed 75,000 people, and the ^{239}Pu-fueled bomb dropped on Nagasaki three days later touched off a violent debate after World War II about the building of the next superweapon – a fusion, or "hydrogen," bomb.

Alumni of the Manhattan project, who had developed the atomic bomb, were divided on the issue. Ernest Lawrence and Edward Teller fought for the construction of the fusion device. J. Robert Oppenheimer and Enrico Fermi argued against it. The decision was made to develop the weapon, and the first artificial fusion reaction occurred when the hydrogen bomb was tested in November 1952.

The history of fusion research is therefore the opposite of fission research. With fission, the reactor came first, and then the bomb was built. With fusion, the bomb was built long before any progress was made toward the construction of a controlled fusion reactor.

More than 40 years after the first hydrogen bomb was exploded, the feasibility of controlled fusion reactions is still open to debate. The reaction that is most likely to fuel the first fusion reactor is the thermonuclear d-t, or deuterium-tritium, reaction. This reaction fuses two isotopes of hydrogen, deuterium (^{2}H) and tritium (^{3}H), to form helium and a neutron.

$${}^{2}_{1}H + {}^{3}_{1}H \rightarrow {}^{4}_{2}He + {}^{1}_{0}n$$

If we consider the implications of this reaction we can begin to understand why it is called a thermonuclear reaction and why it is so difficult to produce in a controlled manner. The d-t reaction requires that we fuse two positively charged particles. This means that we must provide enough energy to overcome the force of repulsion between these particles before fusion can occur. To produce a self-sustaining reaction, we have to provide the particles with enough thermal energy so that they can fuse when they collide.

Each fusion reaction is characterized by a specific ignition temperature, which must be surpassed before the reaction can occur. The d-t reaction has an ignition temperature above 10^{8} K. In a hydrogen bomb, a fission reaction produced by a small atomic bomb is used to heat the contents to the temperature required to initiate fusion. Obtaining the same result in a controlled reaction is much more challenging.

Any substance at temperatures approaching 10^8 K will exist as a completely ionized gas, or plasma. The goals of fusion research at present include the following.

- To achieve the required temperature to ignite the fusion reaction.
- To keep the plasma together at this temperature long enough to get useful amounts of energy out of the thermonuclear fusion reactions.
- To obtain more energy from the thermonuclear reactions than is used to heat the plasma to the ignition temperature.

These are not trivial goals. The only reasonable container for a plasma at 10^8 K is a magnetic field. Both doughnut-shaped (toroidal) and linear magnetic bottles have been proposed as fusion reactors. But reactors that produce high enough temperatures for ignition are not the same as the reactors that have produced long enough confinement times for the plasma to provide useful amounts of energy.

A second approach to a controlled fusion reactor involves hitting fuel pellets containing the proper reagents for the thermonuclear reaction with pulsed beams of laser power. If enough power was delivered, the fuel pellets would collapse upon themselves, or implode, to reach densities several orders of magnitude greater than normal. This could produce a plasma both hot enough and dense enough to initiate fusion reactions.

NUCLEAR SYNTHESIS AND NUCLEAR MEDICINE

NUCLEAR SYNTHESIS

Within the last 60 years, scientists have generated a new story of creation that offers a model for understanding how nuclei are synthesized.

In 1929, Edwin Hubble provided evidence to suggest that our universe is expanding. Between 1946 and 1948, George Gamow and co-workers generated a model which assumed that the primordial substance, or ylem, from which all other matter was created was an extraordinarily hot, dense singularity that exploded in a "Big Bang" and has been expanding ever since.

This model assumes that neutrons in the ylem were transformed into protons by β-decay. Neutrons and protons then combined to form ^{4}He atoms before the temperature and pressure of the fireball decayed to the point at which no further nuclear reactions were possible.

The first-generation stars condensed out of this cloud of hydrogen and helium. As the gas condensed by gravitational attraction, it became warmer. Eventually, the temperature reached 10^7 K, and first-generation, main-sequence stars were born. The temperature at the cores of these stars was high enough to ignite the following thermonuclear reactions.

$$ {}^{1}_{1}H + {}^{1}_{1}H \rightarrow {}^{2}_{1}H + {}^{1}_{+1}e $$

$$ {}^{1}_{1}H + {}^{2}_{1}H \rightarrow {}^{3}_{2}He $$

$$ {}^{3}_{2}He + {}^{3}_{2}He \rightarrow {}^{4}_{2}He + 2\,{}^{1}_{1}H $$

The net result of these reactions is the formation of a helium atom from four protons.

$$ 4\,{}^{1}_{1}H \rightarrow {}^{4}_{2}He + 2\,{}^{0}_{1}e $$

Eventually the heat generated in this reaction was enough to halt the gravitational collapse of the star, which entered a stable period during which the energy generated by this reaction balanced the energy radiated at the surface.

The hydrogen-burning reactions in a main-sequence star are concentrated in the core. When enough hydrogen has been consumed, the core begins to collapse, and the temperature of the core rises above 10^8 K. (The larger the star, the more rapidly it radiates energy from its surface, and the more rapidly it consumes the hydrogen in the core.) As the core collapses, the hydrogen-containing outer shell expands, and the surface of the star cools. Stars that have reached this point in their evolution include the so-called red giants.

Temperatures in the core of these red giants are high enough to ignite further thermonuclear fusion reactions, such as the following.

$$ 3\,{}^{4}_{2}He \rightarrow {}^{12}_{6}C $$

This reaction becomes the principal source of energy in a red giant, although there is undoubtedly some burning of hydrogen to helium in the outer shell of the star. As the amount of ^{12}C in the core increases, further reactions occur to form ^{16}O and ^{20}Ne.

$$ {}^{12}_{6}C + {}^{4}_{2}He \rightarrow {}^{16}_{8}O + \gamma $$

$$ {}^{16}_{8}O + {}^{4}_{2}He \rightarrow {}^{20}_{10}Ne + \gamma $$

Eventually, the helium in the core is exhausted, and the core collapses further, reaching temperatures of 6–7 × 10^8 K. At this point, more complex reactions take place that produce nuclides such as ^{28}Si and ^{32}S.

$$^{12}_{6}C + ^{16}_{8}O \rightarrow ^{28}_{14}Si + \gamma$$

$$^{16}_{8}O + ^{16}_{8}O \rightarrow ^{32}_{16}S + \gamma$$

Further gravitational collapse heats the core to temperatures above 10^9 K, and a complex sequence of reactions takes place to synthesize the nuclei with the highest binding energies, such as Fe and Ni. If the star explodes in a supernova, its contents are ejected across space. Second-generation stars that condense in this region contain not only hydrogen and helium but elements with higher atomic number.

The best estimates of the age of the Milky Way suggest that our galaxy is about 15 billion years old. Our sun and its planets, however, are only 4.6 billion years old. This suggests that the sun is a second-generation star. In such stars, the transformation of hydrogen to helium can be catalyzed by ^{12}C, as shown in the figure below.

$$^{12}_{6}C + ^{1}_{1}H \rightarrow ^{13}_{7}N + \gamma$$

$$^{13}_{7}C + ^{13}_{6}H \rightarrow ^{0}_{7}e$$

$$^{13}_{6}C + ^{1}_{1}H \rightarrow ^{14}_{7}N + \gamma$$

$$^{14}_{7}C + ^{1}_{1}H \rightarrow ^{15}_{8}O + \gamma$$

$$^{15}_{8}O \rightarrow ^{15}_{7}N + ^{0}_{+1}e$$

$$^{15}_{7}N + ^{1}_{1}H \rightarrow ^{12}_{6}C + ^{4}_{2}He$$

Second-generation stars use a different mechanism to synthesis helium than first-generation stars. They use a sequence of reactions, such as the those shown here, that are catalyzed by isotopes heavier than helium.

Two processes can synthesize elements with atomic numbers larger than that of iron. One of them is relatively slow (s-process), the other is very rapid (r-process). Since the only way to synthesize nuclei with atomic numbers larger than iron is by the absorption of neutrons, both the s-process and the r-process result from (n,γ) reactions.

In the s-process, neutrons are captured one at a time to form a neutron-rich nuclide that has enough time to undergo α- or β -decay before another neutron can be absorbed. An example of an s-process sequence of reactions starts with ^{120}Sn. The capture of a neutron produces ^{121}Sn, which undergoes β^--decay. If β^--decay occurs before this nuclide captures another neutron, a stable isotope of antimony is formed. Eventually, ^{121}Sb captures a neutron to produce ^{122}Sb, which is transformed into ^{122}Te by β -decay.

^{122}Te can undergo β -decay to form ^{122}I, or it can capture a neutron to form ^{123}Te. With ^{123}Te, we encounter a series of stable isotopes of tellurium. Thus, neutrons are slowly absorbed, one at a time, until we reach ^{127}Te, which decays to ^{127}I, the most abundant isotope of iodine.

$$ {}^{120}_{50}Sn(n,\gamma)\,{}^{121}_{50}Sn \xrightarrow{\beta^-} {}^{121}_{52}Sb(n,\gamma)\,{}^{122}_{51}Sb $$

$$ \xrightarrow{\beta^-} {}^{122}_{52}Te(5n,5\gamma)\,{}^{127}_{52}Te \xrightarrow{\beta^-} {}^{127}_{52}I $$

This slow process can't account for very heavy nuclides, such as ^{232}Th and ^{238}U, because the lifetimes of the intermediate nuclei with atomic numbers between 83 and 90 are too short for this step-by-step absorption of neutrons to proceed.

Synthesizing appreciable quantities of uranium and thorium requires a rapid process. In the r-process, a number of neutrons are captured in rapid succession, before there is time for α- or β -decay to take place. Achieving an r-process reaction, however, requires a very high neutron flux. (These reactions occur during nuclear explosions, for example.) The neutron flux needed to fuel such reactions is not likely to occur in a normal star. During the moment when a star explodes as a supernova, however, the conditions are ripe for r-process reactions.

The heavier elements on this planet were therefore produced in a series of supernova explosions that occurred in this portion of the galaxy before our solar system condensed.

NUCLEAR MEDICINE

Ever since the first x-ray images were obtained by Roentgen in 1895, ionizing radiation and radio nuclides have played a vital role in medicine. This work has been so fruitful that a separate field known as nuclear medicine has developed. Research in this field focuses on either therapeutic or diagnostic uses of radiation.

There are three standard approaches to fighting cancer: surgery, chemotherapy, and radiation. Surgery, by its very nature, is invasive. Chemotherapy and classic approaches to radiation therapy are not selective. Research in recent years has therefore examined new approaches to radiation therapy that specifically attack tumor cells, without damaging normal tissue. The technique known as boron neutron capture therapy provides an example of this work. Naturally occurring boron consists of two stable isotopes: ^{10}B (19.7%) and ^{11}B (80.3%). ^{10}B absorbs thermal neutrons to form ^{11}B in a nuclear excited state. Although ^{11}B in its nuclear ground state is stable, this excited ^{11}B nuclide undergoes fission to produce ^{7}Li and an α-particle.

$$^{10}_{5}B + ^{1}_{0}n \rightarrow ^{7}_{3}Li + ^{4}_{2}He$$

Because the energy of a thermal neutron is only about 0.025 eV, the neutrons that are not absorbed do relatively little damage to the normal tissue. The α-particle emitted in this reaction has an energy of 2.79 MeV, however, which makes it an extremely lethal form of radiation. The RBE for α-particle radiation is larger than most other particles because this relatively massive particle loses energy very efficiently as it collides with matter. Radiation damage from the α-particle is therefore restricted to the immediate vicinity of the tissue that absorbed the thermal neutron.

Although other common nuclides in living tissue can absorb thermal neutrons, the ability of ^{10}B to absorb thermal neutrons is three orders of magnitude larger than these nuclides.

The units with which this measurement is made can be understood by thinking about the area around the nuclide through which the neutron can pass and still be absorbed. The neutron-capture cross section for a hydrogen atom corresponds to a circle with a radius of about 2×10^{-13} cm.

For a nitrogen atom, the radius of this circle is about 10^{-12} cm. Boron has a neutron-capture cross section that would be described by a circle with a radius of about 2×10^{-9} cm. Virtually all of the neutron capture that occurs is therefore concentrated in the tissue that contains boron.

The potential of boron neutron capture therapy was recognized as early as 1936. Early clinical trials in the late 1950s and early 1960s failed to prolong the lives of patients suffering from brain tumors because the boron compounds available for testing did not concentrate selectively in the tumor cells.

Recent research has discovered boron-labeled compounds that are sufficiently selective for BNCT to become useful as an adjunct to surgery, or in place of surgery with inoperable cases.

Chapter 12

Nuclear Science

Nuclear Science is a subfield of chemistry dealing with radioactivity, nuclear processes and nuclear properties. It is the chemistry of radioactive elements such as the actinides, radium and radon together with the chemistry associated with equipment (such as nuclear reactors) which are designed to perform nuclear processes. This includes the corrosion of surfaces and the behaviour under conditions of both normal and abnormal operation (such as during an accident). An important area is the behaviour of objects and materials after being placed into a waste store or otherwise disposed of.

- The study of the chemical effects resulting from the absorption of radiation within living animals, plants, and other materials. The radiation chemistry controls much of radiation biology as radiation has an effect on living things at the molecular scale, to explain it another way the radiation alters the biochemicals within an organism, the alteration of the biomolecules then changes the chemistry which occurs within the organism, this change in biochemistry then can lead to a biological outcome. As a result nuclear chemistry greatly assists the understanding of medical treatments (such as cancer radiotherapy) and has enabled these treatments to improve.
- The study of the production and use of radioactive sources for a range of processes. These include radiotherapy in medical applications; the use of radioactive tracers within industry, science and the environment; and the use of radiation to modify materials such as polymers.
- The study and use of nuclear processes in *non-radioactive* areas of human activity. For instance, nuclear magnetic resonance (NMR) spectroscopy is commonly used in synthetic organic chemistry and physical chemistry and for structural analysis in macromolecular chemistry.

EARLY HISTORY

After the discovery of X-rays by Wilhelm Röntgen, many scientists began to work on ionizing radiation. One of these was Henri Becquerel, who investigated the relationship between phosphorescence and the blackening of photographic plates. When Becquerel (working in France) discovered that, with no external source of energy, the uranium generated rays which could blacken (or *fog*) the photographic plate, radioactivity was discovered.

Marie Curie (working in Paris) and her husband Pierre Curie isolated two new radioactive elements from uranium ore. They used radiometric methods to identify which stream the radioactivity was in after each chemical separation; they separated the uranium ore into each of the different chemical elements that were known at the time, and measured the radioactivity of each fraction. They then attempted to separate these radioactive fractions further, to isolate a smaller fraction with a higher specific activity (radioactivity divided by mass). In this way, they isolated polonium and radium.

It was noticed in about 1901 that high doses of radiation could cause an injury in humans, Becquerel had carried a sample of radium in his pocket and as a result he suffered a high localised dose which resulted in a radiation burn this injury resulted in the biological properties of radiation being investigated, which in time resulted in the development of medical treatments.

Marie Curie's daughter and her husband were the first to 'create' radioactivity: they bombarded boron with alpha particles to make a proton-rich isotope of nitrogen; this isotope emitted positrons. In addition, they bombarded aluminium and magnesium with neutrons to make new radioisotopes.

Ernest Rutherford, working in Canada and England, showed that radioactivity decay can be described by a simple equation (a linear first degree derivative equation, now called first order kinetics), implying that a given radioactive substance has a characteristic "half life" (the time taken for the amount of radioactivity present in a source to diminish by half).

He also coined the terms alpha, beta and gamma rays, he converted nitrogen into oxygen, and most importantly he supervised the students who did the Geiger-Marsden experiment (gold leaf experiment) which showed that the 'plum pudding model' of the atom was wrong. In the plum pudding model, proposed by J. J. Thomson in 1904, the atom is composed of electrons surrounded by a 'cloud' of positive charge to balance the electrons' negative charge. To Rutherford, the gold foil experiment implied that the positive charge was confined to a very small nucleus leading first to the Rutherford model, and eventually to the Bohr model of the atom, where the positive nucleus is surrounded by the negative electrons.

MAIN AREAS

Radiochemistry is the chemistry of radioactive materials, where radioactive isotopes of elements are used to study the properties and chemical reactions of non-radioactive isotopes (often within radiochemistry the absence of radioactivity leads to a substance being described as being *inactive* as the isotopes are *stable*).

Radiation Chemistry

Radiation chemistry is the study of the chemical effects of radiation on matter; this is very different to radiochemistry as no radioactivity needs to be present in the material which is being chemically changed by the radiation. An example is the conversion of water into hydrogen gas and hydrogen peroxide.

Study of Nuclear Reactions

A combination of radiochemistry and radiation chemistry is used to study nuclear reactions such as fission and fusion. Some early evidence for nuclear fission was the formation of a shortlived radioisotope of barium which was isolated from neutron irradiated uranium (^{139}Ba, with a half-life of 83 minutes and ^{140}Ba, with a half-life of 12.8 days, are major fission products of uranium).

At the time, it was thought that this was a new radium isotope, as it was then standard radiochemical practice to use a barium sulphate carrier precipitate to assist in the isolation of radium. More recently, a combination of radiochemical methods and nuclear physics has been used to try to make new 'superheavy' elements; it is thought that islands of relative stability exist where the nuclides have half-lives of years, thus enabling weighable amounts of the new elements to be isolated.

THE NUCLEAR FUEL CYCLE

The chemistry associated with any part of the nuclear fuel cycle, including nuclear reprocessing. The fuel cycle includes all the operations involved in producing fuel, from mining, ore processing and enrichment to fuel production (*Front end of the cycle*).

It also includes the 'in-pile' behaviour (use of the fuel in a reactor) before the *back end* of the cycle. The *back end* includes the management of the used nuclear fuel in either a cooling pond or dry storage, before it is disposed of into an underground waste store or reprocessed.

Normal and Abnormal Conditions

The nuclear chemistry associated with the nuclear fuel cycle can be divided into two main areas, one area is concerned with operation under the intended conditions while the other area is concerned with maloperation conditions where some alteration from the normal operating conditions has occurred or (*more rarely*) an accident is occurring.

Reprocessing

Law: In the USA it is normal to use fuel once in a power reactor before placing it in a waste store. The long term plan is currently to place the civil used power reactor fuel in a deep store. This policy of not reprocessing was started in March 1977 for nuclear weapons proliferation reasons. The President Jimmy Carter issued a Presidential directive which indefinitely suspended the commercial reprocessing and recycling of plutonium in the USA.

This Presidential directive is likely to have been an attempt by the USA to lead other countries by example, but many other nations continue to reprocess spent nuclear fuels. It is noteworthy that the government under Putin (President of Russia) repealed a law which had banned the import of used nuclear fuel into Russia, this change in Russian law now permits the Russians to offer a reprocessing service for clients outside Russia (In a similar way to that offered by BNFL).

PUREX Chemistry

The current method of choice is to use the PUREX liquid-liquid extraction process which uses a tributyl phosphate/hydrocarbon mixture to extract both uranium and plutonium from nitric acid. This extraction is of the nitrate salts and is classed as being of a solvation mechanism. For example the extraction of plutonium by an extraction agent (S) in a nitrate medium occurs by the following reaction.

$$Pu^{4+}{}_{aq} + 4NO_3{}^{-}{}_{aq} + 2S_{organic} \rightarrow [Pu(NO_3)_4S_2]_{organic}$$

A complex is formed between the metal cation, the nitrates and the tributyl phosphate, and a model compound of a dioxouranium(VI) complex with two nitrates and two triethyl phosphates has been characterised by X-ray crystallography.

When the nitric acid concentration is high the extraction into the organic phase is favoured, and when the nitric acid concentration is low the extraction is reversed (the organic phase is *stripped* of the metal). It is normal to dissolve the used fuel in nitric acid, after the removal of the insoluble matter the uranium and plutonium are extracted from the highly active liquor.

It is normal to then back extract the loaded organic phase to create a *medium active* liquor which contains mostly uranium and plutonium with only small traces of fission products. This medium active aqueous mixture is then extracted again by tributyl phosphate/hydrocarbon to form a new organic phase, the metal bearing organic phase is then stripped of the metals to form an aqueous mixture of only uranium and plutonium.

The two stages of extraction are used to improve the purity of the actinide product, the organic phase used for the first extraction will suffer a far greater dose of radiation. The radiation can degrade the tributyl phosphate into dibutyl hydrogen phosphate. The dibutyl hydrogen phosphate can act as an extraction agent for both the actinides and other metals such as ruthenium.

The dibutyl hydrogen phosphate can make the system behave in a more complex manner as it tends to extract metals by an ion exchange mechanism (extraction favoured by low acid concentration), to reduce the effect of the dibutyl hydrogen phosphate it is common for the used organic phase to be washed with sodium carbonate solution to remove the acidic degradation products of the tributyl phosphate.

New Methods being Considered for Future use

The PUREX process can be modified to make a UREX (URanium EXtraction) process which could be used to save space inside high level nuclear waste disposal sites, such as Yucca Mountain, by removing the uranium which makes up the vast majority of the mass and volume of used fuel and recycling it as reprocessed uranium.

The UREX process is a PUREX process which has been modified to prevent the plutonium being extracted. This can be done by adding a plutonium reductant before the first metal extraction step. In the UREX process, ~99.9% of the Uranium and >95% of Technetium are separated from each other and the other fission products and actinides.

The key is the addition of acetohydroxamic acid (AHA) to the extraction and scrub sections of the process. The addition of AHA greatly diminishes the extractability of Plutonium and Neptunium, providing greater proliferation resistance than with the plutonium extraction stage of the PUREX process.

Adding a second extraction agent, octyl(phenyl)-N, N-dibutyl carbamoylmethyl phosphine oxide(CMPO) in combination with tributylphosphate, (TBP), the PUREX process can be turned into the TRUEX (TRansUranic EXtraction) process this is a process which was invented in the USA by Argonne National Laboratory, and is designed to remove the transuranic metals (Am/Cm) from waste. The idea is that by lowering the alpha activity of the waste, the majority of the waste can then be disposed of with greater ease. In common with PUREX this process operates by a solvation mechanism.

As an alternative to TRUEX, an extraction process using a malondiamide has been devised. The DIAMEX (DIAMideEXtraction) process has the advantage of avoiding the formation of organic waste which contains elements other than Carbon, Hydrogen, Nitrogen, and Oxygen. Such an organic waste can be burned without the formation of acidic gases which could contribute to acid rain.

The DIAMEX process is being worked on in Europe by the French CEA. The process is sufficiently mature that an industrial plant could be constructed with the existing knowledge of the process. In common with PUREX this process operates by a solvation mechanism.

Selective ActiNide EXtraction. As part of the management of minor actinides it has been proposed that the lanthanides and trivalent minor actinides should be removed from the PUREX raffinate by a process such as DIAMEX or TRUEX. In order to allow the actinides such as americium to be either reused in industrial sources or used as fuel the lanthanides must be removed.

The lanthanides has large neutron cross sections and hence they would poison a neutron driven nuclear reaction. To date the extraction system for the SANEX process has not been defined, but currently several different research groups are working towards a process. For instance the French CEA is working on a bis-triaiznyl pyridine (BTP) based process.

Other systems such as the dithiophosphinic acids are being worked on by some other workers.

This is the *UNiversal* EXtraction process which was developed in Russia and the Czech Republic, it is a process designed to remove all of the most troublesome (Sr, Cs and minor actinides) radioisotopes from the raffinates left after the extraction of uranium and plutonium from used nuclear fuel.

The chemistry is based upon the interaction of caesium and strontium with poly ethylene oxide (poly ethylene glycol) and a cobalt carborane anion (known as chlorinated cobalt dicarbollide).

The actinides are extracted by CMPO, and the diluent is a polar aromatic such as nitrobenzene. Other dilents such as *meta*-nitrobenzotrifluoride and phenyl trifluoromethyl sulfone have been suggested as well.

Absorption of Fission Products on Surfaces

Another important area of nuclear chemistry is the study of how fission products interact with surfaces; this is thought to control the rate of release and migration of fission products both from waste containers under normal conditions and from power reactors under accident conditions.

It is interesting to note that, like chromate and molybdate, the $^{99}TcO_4$ anion can react with steel surfaces to form a corrosion resistant layer. In this way, these metaloxo anions act as anodic corrosion inhibitors. The formation of $^{99}TcO_2$ on steel surfaces is one effect which will retard the release of ^{99}Tc from nuclear waste drums and nuclear equipment which has been lost before decontamination (eg submarine reactors lost at sea). This $^{99}TcO_2$ layer renders the steel surface passive, inhibiting the anodic corrosion reaction.

The radioactive nature of technetium makes this corrosion protection impractical in almost all situations. It has also been shown that $^{99}TcO_4$ anions react to form a layer on the surface of activated carbon (charcoal) or aluminium. A short review of the biochemical properties of a series of key long lived radioisotopes can be read on line.

It is important to note that ^{99}Tc in nuclear waste may exist in chemical forms other than the $^{99}TcO_4$ anion, these other forms have different chemical properties. Similarly, the release of iodine-131 in a serious power reactor accident could be retarded by absorption on metal surfaces within the nuclear plant.

SPINOUT AREAS

Some methods first developed within nuclear chemistry and physics have become so widely used within chemistry and other physical sciences that they may be best thought of as separate from *normal* nuclear chemistry.

For example, the isotope effect is used so extensively to investigate chemical mechanisms and the use of cosmogenic isotopes and long-lived unstable isotopes in geology that it is best to consider much of isotopic chemistry as separate from nuclear chemistry.

Kinetics (Use within Mechanistic Chemistry)

The mechanisms of chemical reactions can be investigated by observing how the kinetics of a reaction are changed by making an isotopic modification of a substrate. This is now a standard method in organic chemistry.

Briefly, replacing normal hydrogens (protons) by deuterium within a chemical compound causes the rate of molecular vibration (C-H, N-H and O-H bonds show this) to decrease. This then can lead to a decrease in the reaction rate if the rate-determining step involves breaking a bond between hydrogen and another atom.

Thus, if the reaction changes in rate when protons are replaced by deuteriums, it is reasonable to assume that the breaking of the bond to hydrogen is part of the step which determines the rate.

Uses within Geology, Biology and Forensic Science

Cosmogenic isotopes are formed by the interaction of cosmic rays with the nucleus of an atom. These can be used for dating purposes and for use as natural tracers. In addition, by careful measurement of some ratios of stable isotopes it is possible to obtain new insights into the origin of bullets, ages of ice samples, ages of rocks, and the diet of a person can be identified from a hair or other tissue sample.

Biology

Within living things, isotopic labels (both radioactive and nonradioactive) can be used to probe how the complex web of reactions which makes up the metabolism of an organism converts one substance to another. For instance a green plant uses light energy to convert water and carbon dioxide into glucose by photosynthesis.

If the oxygen in the water is labeled, then the label appears in the oxygen gas formed by the plant and not in the glucose formed in the chloroplasts within the plant cells.

For biochemical and physiological experiments and medical methods, a number of specific isotopes have important applications.

- Stable isotopes have the advantage of not delivering a radiation dose to the system being studied; however, a significant excess of them in the organ or organism might still interfere with its functionality, and the availability of sufficient amounts for whole-animal studies is limited for many isotopes. Measurement is also difficult, and usually requires mass spectroscopy to determine how much of the isotope is present in particular compounds, and there is no means of localizing measurements within the cell.
- H-2 (deuterium), the stable isotope of hydrogen, is a stable tracer, the concentration of which can be measured by mass spectroscopy or NMR. It is incorporated into all cellular structures. Specific deuterated compound can also be produced.
- N-15, the stable isotope of nitrogen, has also been used. It is incorporated mainly into proteins.
- Radioactive isotopes have the advantages of being detectable in very low quantities, in being easily measured by scintillation counting or other radiochemical methods, and in being localizable to particular regions of a cell, and quantifiable by autoradiography. Many compounds with the radioactive atoms in specific positions can be prepared, and are widely available commercially. In high quantities they require precautions to guard the workers from the effects of radiation—and they can easily contaminate laboratory glassware and other equipment.

For some isotopes the half-life is so short that preparation and measurement is difficult.

By organic synthesis it is possible to create a complex molecule with a radioactive label that can be confined to a small area of the molecule. For short-lived isotopes such as ^{11}C, very rapid synthetic methods have been developed to permit the rapid addition of the radioactive isotope to the molecule.

For instance a palladium catalysed carbonylation reaction in a microfluidic device has been used to rapidly form amides and it might be possible to use this method to form radioactive imaging agents for PET imaging.

- ^{3}H, Tritium, the radioisotope of hydrogen, it available at very high specific activities, and compounds with this isotope in particular positions are easily prepared by standard chemical reactions such as hydrogenation of unsaturated precursors. The isotope emits very soft beta radiation, and can be detected by scintillation counting.
- ^{11}C, Carbon-11 can be made using a cyclotron, boron in the form of boric oxide is reacted with protons in a (p,n) reaction. An alternative route is to react ^{10}B with deuterons. By rapid organic synthesis, the ^{11}C compound formed in the cyclotron is converted into the imaging agent which is then used for PET.
- ^{14}C, Carbon-14 can be made (as above), and it is possible to convert the target material into simple inorganic and organic compounds. In most organic synthesis work it is normal to try to create a product out of two approximately equal sized fragments and to use a convergent route, but when a radioactive label is added, it is normal to try to add the label late in the synthesis in the form of a very small fragment to the molecule to enable the radioactivity to be localised in a single group. Late addition of the label also reduces the number of synthetic stages where radioactive material is used.
- ^{18}F, fluorine-18 can be made by the reaction of neon with deuterons, ^{20}Ne reacts in a (d,^{4}He) reaction. It is normal to use neon gas with a trace of stable fluorine ($^{19}F_2$). The $^{19}F_2$ acts as a carrier which increases the yield of radioactivity from the cyclotron target by reducing the amount of radioactivity lost by absorption on surfaces. However, this reduction in loss is at the cost of the specific activity of the final product.

Nuclear Magnetic Resonance (NMR)

NMR spectroscopy uses the net spin of nuclei in a substances upon energy absorption to identify molecules. This has now become a standard spectroscopic tool within synthetic chemistry. One major use of NMR is to determine the bond connectivity within an organic molecule.

NMR imaging also uses the net spin of nuclei (commonly protons) for imaging. This is widely used for diagnostic purposes in medicine, and can provide detailed images of the inside of a person without inflicting any radiation upon them. In a medical setting, NMR is often known simply as "magnetic resonance" imaging, as the word 'nuclear' has negative connotations for many people.

NUCLEAR CHEMISTRY

Nuclear Chemistry deals with the nuclei of atoms breaking apart. Atoms are continually undergoing decay. When studying nuclear chemistry, there is a typical format used to represent specific isotopes.

Mass Number

$$\swarrow$$

$$^{238}_{92}\text{U} \leftarrow \text{Element}$$

$$\nwarrow$$

Charge or Atomic Number

Nuclear equations are typically written in the format shown below. There are 5 different types of radioactive decay.

1. *Alpha decay follows the form*:

$$^{X}_{Z}A \rightarrow \underset{\text{Alpha Particle}}{\underset{\downarrow}{^{4}_{2}He}} + ^{X-4}_{Z-2}B$$

Where A is the parent isotope (the atom being broken apart) B is the daughter isotope or the isotope formed. When an element is broken down in alpha decay it looces two neutrons and two (2) protons. This means hat the name of the element will change as well, moving back two (2) places on the periodic table. Alpha decay is is not very penetrating because the He atoms capture electrons before traveling very far. However it is very damaging because the alpha particles can knock atoms off of molecules.Alpha decay is the most common in elements with an atomic number greater than 83.

2. *Beta negative decay follows the form*:

$$^{X}_{Z}A \rightarrow \underset{\text{Beta Particle}}{\underset{\downarrow}{^{0}_{-1}e^{-1}}} + ^{X}_{Z-1}B$$

The beta emission increases the atomic number by one (1) by adding one (1) proton. At the same time, one (1) neutron is lost so the mass of the daughter isotope is the same as the parent isotope. Beta negative decay is more penetrating than alpha decay because the particles are smaller, but less penetrating than gamma decay. Beta electrons can penetrate through about one (1) cm of flesh before they are brought to a halt because of electrostatic forces. Beta decay is most common in elements with a high neutron to proton ratio.

3. *Gamma decay follows the form:*

$$^{X}_{Z}A \rightarrow \underset{\downarrow}{^{0}_{0}\gamma} + ^{X}_{Z}B$$

Gamma Particle

In gamma emission, neither the atomic number or the mass number is changed. A high energy gamma ray is given off when the parent isotope falls into a lower energy state. Gamma radiation is the most penetrating of all. These photons can pass through the body and cause damage by ionizing all the molecules in their way.

4. *Positron emission (also called Beta positive decay) follows the form:*

$$^{X}_{Z}A \rightarrow \underset{\downarrow}{^{0}_{1}e} + ^{X}_{Z-1}B$$

Positron Particle

In this reaction a positron is emitted. A positron is exactly like an electron in mass and charge force except with a positive charge. It is formed when a proton breaks into a neutron with mass and no charge and this positron with no mass and the positive charge. Positron emission is most common in lighter elements with a low neutron to proton ratio.

5. *Electron capture follows the form:*

$$^{X}_{Z}A + \underset{\downarrow}{^{0}_{-1}e} + ^{X}_{Z-1}B$$

Electron Particle

In this reaction a nucleus captures one (1) of its own atom's inner shell electrons which reduces the atomic number by one. This captured electron joins with a proton in the nucleus to form a neutron. Electron capture is common in larger elements with a low neutron to proton ratio.

All elements with an atomic number over 83 are considered radioactive. Radioactivity can be measured using a geiger counter, a cylinder containing a low-pressure gas and two (2) electrodes. Radiation ionizes the atoms in the cylinder and allows current to flow between the electrodes. The half life of a radioactive substance is the time required for half of the initial number of nuclei to disintegrate. The decay rate expresses the speed at which a substance disintegrates. The following equation represents the relationship between the number of nuclei remaining, N, the number of nuclei initially present, N_O, the rate of decay, k, and the amount of time, t.

$$\text{In}\frac{N}{N_0} = -kt$$

The relationship between the half-life of a radioactive substance and k, the rate at which it decays can also be found.

$$t_{1/2} \frac{.693}{k}$$

By using these equations, it is possible to calculate how much of a nuclear substance will be left after a certain time and how much of a substance originally existed. A common example is isotopic dating in which the ages of archeological artifacts are determined by measuring the activity of the isotopes.

RADIOACTIVITY: HISTORICAL FIGURES

ACCESS EXCELLENCE CLASSIC COLLECTION

This article will focus on the efforts of four scientists: Wilhelm Conrad Roentgen, Antoine Henri Becquerel, Marie Sklodowska Curie, and Ernest Rutherford.

It emphasizes their contributions to the elucidation of radioactivity and the "key" experiments they performed pertaining to their discoveries. The biographies and photographs are adapted from The Health Physics Society Centennial Calendar by permission of the Health Physics Society.

Wilhelm Conrad Roentgen

On November 8, 1895, at the University of Wurzburg, Wilhelm Roentgen's attention was drawn to a glowing fluorescent screen on a nearby table.

Roentgen immediately determined that the fluorescence was caused by invisible rays originating from the partially evacuated glass Hittorf-Crookes tube he was using to study cathode rays (i.e., electrons). Surprisingly, these mysterious rays penetrated the opaque black paper wrapped around the tube. Roentgen had discovered X rays, a momentous event that instantly revolutionized the field of physics and medicine.

However, prior to his first formal correspondence to the University Physical-Medical Society, Roentgen spent two months thoroughly investigating the properties of X rays. Silvanus Thompson complained that Roentgen left "little for others to do beyond elaborating his work." For his discovery, Roentgen received the first Nobel Prize in physics in 1901. When later asked what his thoughts were at the moment of his discovery, he replied "I didn't think, I investigated. "It was the crowning achievement in a career beset by more than its share of difficulties.

As a student in Holland, Roentgen was expelled from the Utrecht Technical School for a prank committed by another student. Even after receiving a doctorate, his lack of a diploma initially prevented him from obtaining a position at the University of Wurzburg. He even was accused of having stolen the discovery of X rays by those who failed to observe them.

Nevertheless, Roentgen was a brilliant experimentalist who never sought honors or financial profit for his research. He rejected a title (i.e., von Roentgen) that would have provided entry into the German nobility, and donated the money he received from the Nobel Prize to his University.

Roentgen did accept the honorary degree of Doctor of Medicine offered to him by the medical faculty of his own University of Wurzburg. However, he refused to take out any patents in order that the world could freely benefit from his work. At the time of his death, Roentgen was nearly bankrupt from the inflation that followed World War I.

Antoine Henri Becquerel

Henri Becquerel was born into a family of scientists. His grandfather had made important contributions in the field of electrochemistry while his father had investigated the phenomena of fluorescence and phosphorescence. Becquerel not only inherited their interest in science, he also inherited the minerals and compounds studied by his father.

And so, upon learning how Wilhelm Roentgen discovered X rays from the fluorescence they produced, Becquerel had a ready source of fluorescent materials with which to pursue his own investigations of these mysterious rays.

The material Becquerel chose to work with was potassium uranyl sulfate, $K_2UO_2(SO_4)_2$, which he exposed to sunlight and placed on photographic plates wrapped in black paper. When developed, the plates revealed an image of the uranium crystals. Becquerel concluded "that the phosphorescent substance in question emits radiation which penetrates paper opaque to light." Initially he believed that the sun's energy was being absorbed by the uranium which then emitted X rays.

Further investigation, on the 26th and 27th of February, was delayed because the skies over Paris were overcast and the uranium-covered plates Becquerel intended to expose to the sun were returned to a drawer. On the first of March, he developed the photographic plates expecting only faint images to appear. To his surprise, the images were clear and strong. This meant that the uranium emitted radiation without an external source of energy such as the sun. Becquerel had discovered radioactivity, the spontaneous emission of radiation by a material.

Later, Becquerel demonstrated that the radiation emitted by uranium shared certain characteristics with X rays but, unlike X rays, could be deflected by a magnetic field and therefore must consist of charged particles. For his discovery of radioactivity, Becquerel was awarded the 1903 Nobel Prize for physics.

Marie Curie

By the time he met Marie Sklodowska, Pierre Curie had already established an impressive reputation. In 1880, he and his brother Jacques had discovered piezoelectricity whereby physical pressure applied to a crystal resulted in the creation of an electric potential.

He also had made important investigations into the phenomenon of magnetism including the identification of a temperature, the curie point, above which a material's magnetic properties disappear. However, shortly after his marriage to Marie in 1895, Pierre subjugated his research to her interests.

Together, they began investigating the phenomenon of radioactivity recently discovered in uranium ore. Although the phenomenon was discovered by Henri Becquerel, the term radioactivity was coined by Marie. After chemical extraction of uranium from the ore, Marie noted the residual material to be more "active" than the pure uranium.

She concluded that the ore contained, in addition to uranium, new elements that were also radioactive. This led to their discoveries of the elements of polonium and radium, but it took four more years of processing tons of ore under oppressive conditions to isolate enough of each element to determine its chemical properties.

For their work on radioactivity, the Curies were awarded the 1903 Nobel Prize in physics. Tragically, Pierre was killed three years later in an accident while crossing a street in a rainstorm. Pierre's teaching position at the Sorbonne was given to Marie. Never before had a woman taught there in its 650 year history!

Her first lecture began with the very sentence her husband had used to finish his last. In his honour, the 1910 Radiology Congress chose the curie as the basic unit of radioactivity: the quantity of radon in equilibrium with one gram of radium (current definition: 1 Ci = 3.7×10^{10} dps). A year later, Marie was awarded the Nobel Prize in chemistry for her discoveries of radium and polonium, thus becoming the first person to receive two Nobel Prizes.

For the remainder of her life she tirelessly investigated and promoted the use if radium as a treatment for cancer. Marie Curie died July 4, 1934, overtaken by pernicious anemia no doubt caused by years of overwork and radiation exposure.

Ernest Rutherford

Ernest Rutherford is considered the father of nuclear physics. Indeed, it could be said that Rutherford invented the very language to describe the theoretical concepts of the atom and the phenomenon of radioactivity. Particles named and characterized by him include the alpha particle, beta particle and proton.

Even the neutron, discovered by James Chadwick, owes its name to Rutherford. The exponential equation used to calculate the decay of radioactive substances was first employed for that purpose by Rutherford and he was the first to elucidate the related concepts of the half-life and decay constant.

With Frederick Soddy at McGill University, Rutherford showed that elements such as uranium and thorium became different elements (i.e., transmuted) through the process of radioactive decay. At the time, such an incredible idea was not to be mentioned in polite company: it belonged to the realm of alchemy, not science.

For this work, Rutherford won the 1908 Nobel Prize in chemistry. In 1909, now at the University of Manchester, Rutherford was bombarding a thin gold foil with alpha particles when he noticed that although almost all of them went through the gold, one in eight thousand would "bounce" (i.e., scatter) back. The amazed Rutherford commented that it was "as if you fired a 15-inch naval shell at a piece of tissue paper and the shell came right back and hit you."

From this simple observation, Rutherford concluded that the atom's mass must be concentrated in a small positively-charged nucleus while the electrons inhabit the farthest reaches of the atom. Although this planetary model of the atom has been greatly refined over the years, it remains as valid today as when it was originally formulated by Rutherford.

In 1919, Rutherford returned to Cambridge to become director of the Cavendish laboratory where he had previously done his graduate work under J.J. Thomson. It was here that he made his final major achievement, the artificial alteration of nuclear and atomic structure. By bombarding nitrogen with alpha particles, Rutherford demonstrated the production of a different element, oxygen. "Playing with marbles" is what he called; the newspapers reported that Rutherford had "split the atom." After his death in 1937, Rutherford's remains were buried in Westminster Abbey near those of Sir Isaac Newton.

Chapter 13

Systems of Nuclear Instrumentation

In a car you need to know how fast you are driving so you do not exceed the speed limit. In a nuclear power plant, the operator must know how much neutron power is being produced so that he or she does not exceed the regulatory power limit. Neutron Power level (or power level, for short) can actually span about 11 orders of magnitude (11 powers of 10) from the lowest power level to full power. No one type of detector can be used. In many plants, 3 levels, and methods, of detection are used. These levels are called:

- Source Range (usually a proportional detector)
- Intermediate Range (usually a compensated ion chamber)
- Power Range (usually an uncompensated ion chamber)

The neutron detectors, used by the operator, may be installed inside or outside the reactor, depending on the type of reactor. If detectors normally used by the operator are outside the reactor, these detectors must be periodically recalibrated with detectors that are either installed in the reactor core or that pass through the reactor core.

The neutron detectors are typically cylindrical with a central wire. The wire and the outside of the detector have voltage differences between 10 and 2500 volts dc. The detector often has a gas that will interact with the neutron to produce a charged particle. The charged particles then are drawn to the appropriate central wire or outside of the detector. These events produce either a single pulse or a current depending on the type of detector. Gases used in neutron detectors usually include boron or uranium. When a neutron reacts with boron, it produces 2 charged particles - an alpha particle (helium nucleus ${}_2He^4$ with a +2 charge). When a neutron reacts with the uranium it fissions creating 2 fission products that are both highly positively charged.

CURRENT TRENDS IN NUCLEAR INSTRUMENTATION IN DIAGNOSTIC NUCLEAR MEDICINE

In nuclear medicine imaging the use of multi-detector systems for total-body, brain and heart scanning have recently gained increasing popularity. The systems with 2 or 3 Anger type gamma cameras 180°, 120° or 90° apart with small or large filed of view are most popular. These systems have considerably improved the sensitivity, resolution and the scanning time comparing to the single camera systems.

PET imaging with rings of small detectors in multi-slice configuration with lead or tungsten septa between slices (2-D) or without septa (3-D) has been successfully implemented using FDG in a limited axial region and most recently also for total-body tomography.

The biggest improvement in spatial resolution of PET in the past was gained mainly by the reducing the size of the crystals which currently resolve the structures to 5 mm in size. There is a strong development of the imaging systems with a large number of tiny crystals made by the newly developed high dense and fast responding materials.

Principles of Image Detection

Generally three basic methods of nuclear medicine image formation are actual: 1. single large scintillation crystal with a large number of photo multiplier tubes (PMT), 2. large number of tiny scintillation crystals with position sensitive photo-multiplier tube (PSPMT) and 3. large semiconductor crystal with array of tiny "n-p" sensitive areas.

In the first method the gamma ray strikes a large circular or rectangle thin crystal and the induced scintillation light is distributed between the PMTs according to their viewing spatial angles. The x and y positions of PMTs are weighted by their electric signal responses from all PMTs and the X and Y coordinates of the scintillation cloud striking the array of PMT and the corresponding energy is computed (Anger type of gamma camera).

The intrinsic spatial resolution of the imaging device strongly depends on the crystal thickness, slightly less on the size / number / shape of PMT and the position circuit. The sensitivity of the system oppositely to the spatial resolution increases by the crystal thickness. The energy and time resolution depend mainly on the crystal material. Nowadays nearly all planar gamma cameras are of this kind.

In the second method the gamma ray is absorbed in a tiny crystal and all the induced scintillation light is collected by a small area of position sensitive photo multiplier tube which converts the incident light in a very thin layer into a charge or current which is then converted to digital E (energy) signal. Each small sensitive area of PSPMT provides also corresponding spatial coordinates X and Y for the particular exposed crystal. The spatial resolution strongly depends on the size of the crystals and on the thickness and material of the septa.

Each crystal represents a pixel in the final digital image. Thinner and longer the crystal and thinner the septa better is the resolving power of the imaging system. Still, there is a limitation in the cross sectional size of the crystal.

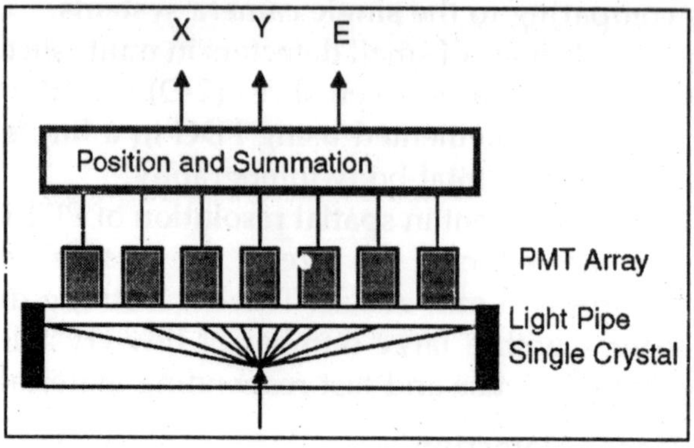

Fig. Imaging Detector with Single Large Scintillation Crystal with Set of PMTs.

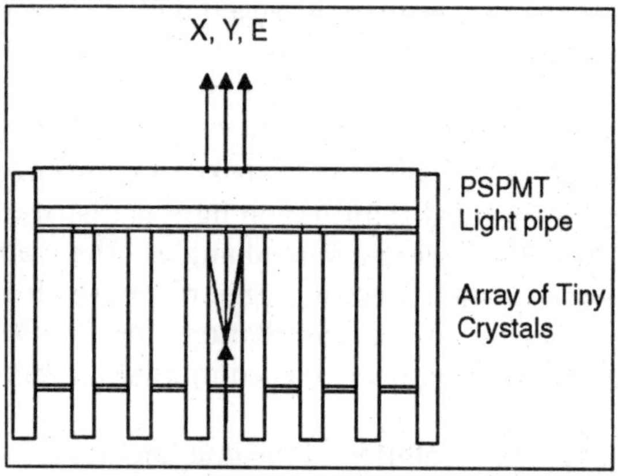

Fig. Imaging Detector with Array of Tiny Scintillation Crystals with PSPMT.

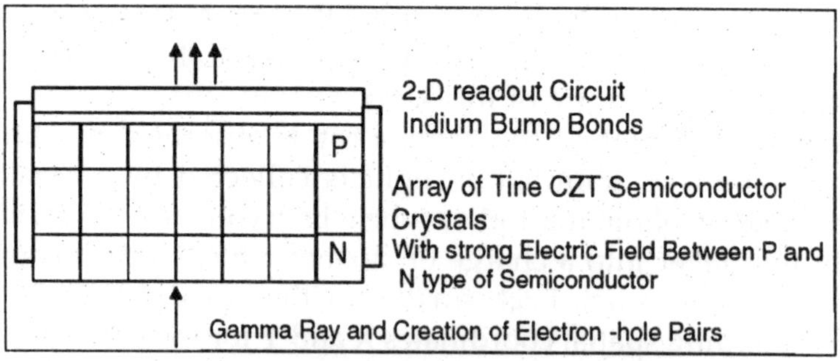

Fig. Imaging Semiconductor Array Detector.

If gamma ray strikes the crystal too close to the reflector cover or to the optical fibre (crystal is inside optical fibre) then some of the ionization doesn't contribute to the scintillation light and therefore the energy signal is smaller. Because of this, a definite volume close to the edge, is not useful and is treated as scattered.

An estimate is given for the circular shape of the crystal and for two popular scintillation materials (NaI and BGO). For approximately 20 % of absorbed gamma rays, the reduction of the scintillation light will be present. Some of the signals coming from these 20 % will still be included in the lower part of the photopeak but some will be lost.

The conclusion is that there is no meaning of using thinner sized crystals than 1-2 mm depending on the material (for NaI crystal this size is limited to 2 mm and for BGO crystal 1 mm). The sensitivity increases by the crystal length, density and cross sectional size. The energy resolution is improved by more efficient PSPMT and more effective collecting of scintillation light in crystal.

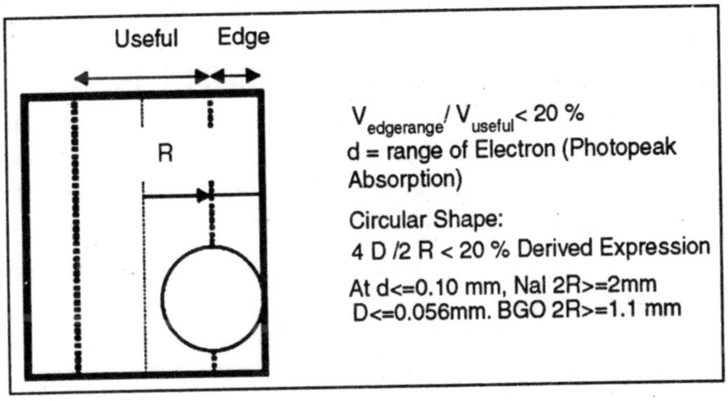

Fig. Estimate of the Limit for the Crystal Cross Eectional Size.

In the third method the incident gamma ray is absorbed in the region of "p-n junction" region of the semiconductor crystal and a large number of electron-hole pairs are created. Their number (approximately 3 – 5 eV/ electron-hole pair is spent on average) is proportional to the energy of the gamma ray and is nearly ten times greater than the quantity of the scintillation light (approximately 30 eV per ionization).

For the same factor the energy resolution is better. The efficiency of the late developed CZT crystal is even better than for NaI. The intrinsic spatial resolution of the CZT gamma camera is considerably better and is about 2 – 3 mm for 140 keV compared to 3 – 4 mm for NaI gamma camera. On the other hand the collection time for electron-hole pairs is at least 100 times shorter than is the decay time for scintillation light in NaI crystal and therefore the increased count rate can be achieved (250.000 counts/s).

The weight of such imaging system is 100 times lower and the CZT gamma camera is easily portable to emergency departments or elsewhere. It is expected that the price for such gamma camera will also be much lower because of the less complicated production.

New Materials for Detection Crystals

Some new detector materials were developed recently which promise a considerable improvement of nuclear medicine imaging devices. The LSO is intrinsically radioactive and is not useful for SPET but can be used for PET (coincidence measurement excludes the single decay and absorption of gamma ray inside LSO crystal).

Very promising is the use of LSO and YSO in so-called phoswhich detector where the YSO crystal (1 - 2 cm) is in front and the LSO crystal (1 - 2 cm) is optically coupled to the YSO. The YSO is used for attenuation of low energy (in the range of 100 keV) and LSO serves as the light pipe and for attenuation of high energy gamma rays. The induced scintillation signals from both crystals can be separated because of the difference in the decay times.

The BGO is nearly exclusively used for PET but will probably be replaced by this phoswhich detector. One of the most interesting detectors becomes semiconductor CZT (cadmium zinc telluride) which has even better stopping power for 140 keV gamma rays than NaI (Tl) but much better energy resolution (for factor of 10). The array with large number of very small sensitive areas can be build so that each of these areas is a pixel in the digital image.

SPET

The biggest advance in the SPET was the implementation of several detector heads which drastically improved the system sensitivity. For cardiac SPET the use of two heads at 90o and the whole-body bone SPET or scanning at 180o shortens the acquisition time or increases the acquisition counts two times.

The improvement of the gamma camera features was mainly due to the development of so-called digital head electronics which had replaced the old analog position circuit by the digital one. The output from each PMT is digitized and then the spatial coordinates are computed. All corrections for non-linearity, spatial and energy non-uniformity can be performed on-line by the use of special fast processors.

In the future the probable development of SPET will be building the tomographic system from the arrangement of several modular multi-crystal detectors which will introduce even greater flexibility than with two or three big planar gamma cameras at different angular setup.

Each module will probably be an array of very tiny crystals from YSO and LSO or semiconductor CZT detector. Another possible approach in modular design of SPET will be in a specialization for some organs (i.e. thyroid and cardiac tomograph needs relatively small sized detector's modules) and possible much lower prices for small SPET systems.

PET

The latest improvements are mainly due to the development of the so called 3–D tomographs which don't use septa between planes with rings of crystals. By omitting the septa the sensitivity is increased by ten times and the amount of scattered photons only approximately 3o %. In such configuration of several thousands tiny crystals all possible coincidence events between any two crystals are used in the reconstruction algorithm.

The system works in a true 3-D mode. Another possibility of PET is the use of dual-head SPET system with or without collimator (high-speed electronics is essential) in a coincidence mode. But this type of PET is considerably less sensitive but is interesting for performing both PET and SPET studies. A much better sensitivity of the system is expected in future from the new generation of the PET.

It will be of extreme importance in the imaging of specific biochemical bindings, such as receptor binding. In this applications a small amount of the injected radioactivity is collected by a target organ (usually less than 1 %).

Surgical Gamma Probe

This application of radioactivity tracing becomes very important in surgery for identifying the regional metastases. Currently interesting clinical field where the small detector probe is of great importance is lymph node dissection of the axilla or regional nodes in the breast cancer and in some melanoma patients.

The role of the surgical gamma probe is to localise the sentinel node transcutaneously and intra-operatively. To meet a high sensitivity, good spatial and spectral resolution and appropriate ergonomic characteristics several different commercially available probes were evaluated. It was found that CZT probe was the most appropriate for low energies (140 keV from 99mTc) and the NaI probe for high energy (364 keV 131I).

Nuclear Medicine instrumentation is passing vigorous development last years and most likely in the near future as well. New detection materials with much better physical characteristics than the standard NaI concerning the stopping power, energy resolution, fragility, decay time, light output, and density will most likely replace the NaI.

It is expected that new imaging devices with several thousands of tiny crystals or semiconductor array of small position sensitive areas will improve the sensitivity and specificity of clinical studies. At the same time the small surgical probes made from these materials are also becoming very popular in surgery tracing the regional metastases.

MAGNETIC RESONANCE IMAGING

Medical Imaging

Medical imaging refers to the techniques and processes used to create images of the human body (or parts thereof) for clinical purposes (medical procedures seeking to reveal, diagnose or examine disease) or medical science (including the study of normal anatomy and function).

As a discipline and in its widest sense, it is part of biological imaging and incorporates radiology (in the wider sense), radiological sciences, endoscopy, (medical) thermography, medical photography and microscopy (e.g. for human pathological investigations).

Measurement and recording techniques which are not primarily designed to produce images, such as electroencephalography (EEG) and magnetoencephalography (MEG) and others, but which produce data susceptible to be represented as maps (i.e. containing positional information), can be seen as forms of medical imaging.

In the clinical context, medical imaging is generally equated to Radiology or "clinical imaging" and the medical practitioner responsible for interpreting (and sometimes acquiring) the images is a radiologist. Diagnostic radiography designates the technical aspects of medical imaging and in particular the acquisition of medical images. The *radiographer* or *radiologic technologist* is usually responsible for acquiring medical images of diagnostic quality, although some radiological interventions are performed by radiologists.

As a field of scientific investigation, medical imaging constitutes a sub-discipline of biomedical engineering, medical physics or medicine depending on the context: Research and development in the area of instrumentation, image acquisition (e.g. radiography), modelling and quantification are usually the preserve of biomedical engineering, medical physics and computer science.

Research into the application and interpretation of medical images is usually the preserve of radiology and the medical sub-discipline relevant to medical condition or area of medical science (neuroscience, cardiology, psychiatry, psychology, etc) under investigation. Many of the techniques developed for medical imaging also have scientific and industrial applications.

Medical imaging is often perceived to designate the set of techniques that noninvasively produce images of the internal aspect of the body. In this restricted sense, medical imaging can be seen as the solution of mathematical inverse problems. This means that cause (the properties of living tissue) is inferred from effect (the observed signal). In the case of ultrasonography the probe consists of ultrasonic pressure waves and echoes inside the tissue show the internal structure. In the case of projection radiography, the probe is X-ray radiation which is absorbed at different rates in different tissue types such as bone, muscle and fat.

MODERN IMAGING TECHNOLOGY

Fluoroscopy

Fluoroscopy produces real-time images of internal structures of the body in a similar fashion to radiography, but employs a constant input of x rays. Contrast media, such as barium, iodine, and air are used to visualize internal organs as they work. Fluoroscopy is also used in image-guided procedures when constant feedback during a procedure is required.

Magnetic Resonance Imaging (MRI)

A Magnetic Resonance Imaging instrument (MRI scanner) uses powerful magnets to polarise and excite hydrogen nuclei (single proton) in water molecules in human tissue, producing a detectable signal which is spatially encoded resulting in images of the body.

In brief, MRI involves the use of three kinds of electromagnetic field: a very strong (of the order of units of teslas) static magnetic field to polarize the hydrogen nuclei, called the static field; a weaker time-varying (of the order of 1 kHz) for spatial encoding, called the gradient field(s); and a weak radio-frequency (RF) field for manipulation of the hydrogen nuclei to produce measurable signals, collected through an RF antenna.

Like CT, MRI traditionally creates a 2D image of a thin "slice" of the body and is therefore considered a tomographic imaging technique. Modern MRI instruments are capable of producing images in the form of 3D blocks, which may be considered a generalisation of the single-slice, tomographic, concept. Unlike CT, MRI does not involve the use of ionizing radiation and is therefore not associated with the same health hazards; for example there are no known long term effects of exposure to strong static fields and therefore there is no limit on the number of scans to which an individual can be subjected, in contrast with X-ray and CT.

However, there are well identified health risks associated with tissue heating from exposure to the RF field and the presence of implanted devices in the body, such as pace makers. These risks are strictly controlled as part of the design of the instrument and the scanning protocols used.

CT and MRI being sensitive to different properties of the tissue, the appearance of the images obtained with the two techniques differ markedly. In CT, X-rays must be blocked by some form of dense tissue to create an image, therefore the image quality when looking at soft tissues will be poor.

While any nucleus with a net nuclear spin can be used, the proton of the hydrogen atom remains the most widely used, especially in the clinical setting, since it is so ubiquitous and returns much signal. This nucleus, present in water molecules, allows excellent soft-tissue contrast.

MRI, or "NMR imaging" as it was originally known, has only been in use since the early 1980s. Effects from long term, or repeated exposure, to the intense static magnetic field are not known.

Nuclear Medicine

Images from gamma cameras are used in Nuclear Medicine to detect regions of biological activity that are often associated with diseases. A short lived isotope, such as ^{131}I is administered to the patient. These isotopes are more readily absorbed by biologically active regions of the body, such as tumors or fracture points in bones.

Positron Emission Tomography (PET)

Positron emission tomography is primarily used to detect diseases of the brain and heart. Similarly to nuclear medicine, a short-lived isotope, such as ^{18}F, is incorporated into a substance used by the body such as glucose which is absorbed by the tumor of interest.

PET scans are often viewed along side computed tomography scans, which can be performed on the same equipment without moving the patient. This allows the tumors detected by the PET scan to be viewed next to the rest of the patient's anatomy detected by the CT scan.

Projection Radiography

Radiographs, more commonly known as x-rays, are often used to determine the type and extent of a fracture as well as for detecting pathological changes in the lungs.

With the use of radio-opaque contrast media, such as barium, they can also be used to visualize the structure of the stomach and intestines - this can help diagnose ulcers or certain types of colon cancer.

Photoacoustic Imaging

Photoacoustic imaging is a recently developed hybrid biomedical imaging modality based on the photoacoustic effect. It combines the advantages of optical absorption contrast with ultrasonic spatial resolution for deep imaging in (optical) diffusive or quasi-diffusive regime.

Recent studies have shown that photoacoustic imaging can be used in vivo for tumor angiogenesis monitoring, blood oxygenation mapping, functional brain imaging, and skin melanoma detection etc.

TOMOGRAPHY

Tomography is the method of imaging a single plane, or slice, of an object resulting in a tomogram. There are several forms of tomography:

Linear Tomography

This is the most basic form of tomography. The X-ray tube moved from point "A" to point "B" above the patient, while the cassette holder (or "bucky") moves simultaneously under the patient from point "B" to point "A."

The fulcrum, or pivot point, is set to the area of interest. In this manner, the points above and below the focal plane are blurred out, just as the background is blurred when panning a camera during exposure. No longer carried out and replaced by computed tomography.

Poly Tomography

This was a complex form of tomography. With this technique, a number of geometrical movements were programmed, such as hypocycloidic, circular and elliptical. Philips Medical Systems produced one such device called the 'Polytome.' No longer carried out, replaced by computed tomography.

Zonography

This is a variant of linear tomography, where a limited arc of movement is used. It is still used in some centres for visualising the kidney during an intravenous urogram (IVU)

Orthopantomography

The only common tomographic examination in use. This makes use of a complex movement to allow the radiographic examination of the mandible, as if it were a flat bone. It is often referred to as a "Panaray", but this is incorrect, as it is a trademark of a specific company's equipment

Computed Tomography (CT)

A CT scan, also known as a CAT scan, is a helical tomography (latest generation), which traditionally produces a 2D image of the structures in a thin section of the body. It uses X-rays. It has a greater ionizing radiation dose burden than projection radiography; repeated scans should be limited.

CLINICAL IMAGING OR BIOLOGICAL IMAGING TECHNIQUES

Electron Microscopy

The electron microscope is a microscope that can magnify very small details with high resolving power due to the use of electrons as the source of illumination, magnifying at levels up to 2,000,000 times.

Electron microscopy is employed in anatomic pathology to identify organelles within the cells. Its usefulness has been greatly reduced by immunhistochemistry but it is still irreplaceable for the diagnosis of kidney disease, identification of immotile cilia syndrome and many other tasks

Creation of Three-dimensional Images

Recently, techniques have been developed to enable CT, MRI and ultrasound scanning software to produce 3D images for the physician. Traditionally CT and MRI scans produced 2D static output on film. To produce 3D images, many scans are made, then combined by computers to produce a 3D model, which can then be manipulated by the physician. 3D ultrasounds are produced using a somewhat similar technique.

With the ability to visualize important structures in great detail, 3D visualization methods are a valuable resource for the diagnosis and surgical treatment of many pathologies. It was a key resource for the famous, but ultimately unsuccessful attempt by Singaporean surgeons to separate Iranian twins Ladan and Laleh Bijani in 2003. The 3D equipment was used previously for similar operations with great success.

Other proposed or developed techniques include:

- Diffuse optical tomography
- Elastography
- Electrical impedance tomography
- Optoacoustic imaging
- Ophthalmology
  - *A*-scan
  - *B*-scan
  - Corneal topography
  - Heidelberg retinal tomography
  - Optical coherence tomography
  - Scanning laser ophthalmoscopy

Some of these techniques are still at a research stage and not yet used in clinical routines.

NON-DIAGNOSTIC IMAGING

Neuroimaging has also been used in experimental circumstances to allow people (especially disabled persons) to control outside devices, acting as a brain computer interface.

Medical Imaging Service

This is a specialized area of medical equipment service and repair, which is separate from the biomedical field, although a hospital with their own service group may include them in the biomed department.

At one time, there were only two ways to receive training for this field. One was to learn it in the military, and the other was on-the-job training (OJT) from the manufacturer. But since the 1980s several independent training centers have been started. One such school is RSTI.

There are different means of employment in this occupation. Working for the manufacturer's field service department (OEM), working for a hospital (in-house), and working for an independent (outside, or independent provider). The most stable positions are with the OEM or hospital, as you can remain current through on-going training, and the two have good working relationships.

The OEM service engineer can expect to spend a lot of time driving from one site to another during the work day, and working non-standard hours. They will install, remove, diagnose, repair, calibrate, perform preventive maintenance, and interface equipment, all while ensuring good customer relations. You may also be required to do yearly testing of the radiation sources for Federal and State compliance.

The in-house person is employed by the hospital. With larger medical facilities, travel between the hospitals' other locations may be required to perform the required services. You may also be required to do yearly testing of the radiation sources for compliance. The OEM or independent will provide installation of purchased equipment, and can be used for back-up service.

An independent is typically someone who has left an OEM, and started their own service business. Staying up-to-date as an independent can be difficult and expensive, as the OEM is usually reluctant to provide training. However, non-OEM training facilities are available, such as the aforementioned RSTI. Competition for service can be aggressive, with OEM's giving hospitals or clinics a reduction in equipment purchase price if they retain some form of OEM service.

The independent may also sell and install refurbished equipment, or de-install equipment. They will repair, calibrate, and perform preventative maintenance. Because many of the tasks associated with imaging service require expensive, specialized equipment, there may be a financial limit to the independent. Typical equipment used routinely are a Storage Oscilloscope and multimeter (if servicing old vacuum-state equipment, a VOM would be helpful). Additional equipment: Keithley dosimeter, mAs meter, Biddle contact tachometer, light to radiation template, etc.

Magnetic resonance imaging (MRI) is primarily used in medical imaging to visualize the structure and function of the body. It provides detailed images of the body in any plane. MR has much greater soft tissue contrast than Computed tomography (CT) making it especially useful in neurological, musculoskeletal, cardiovascular and oncolological diseases.

Unlike CT it uses no ionizing radiation. The scanner creates a powerful magnetic field which aligns the magnetization of hydrogen atoms in the body. Radio waves are used to alter the alignment of this magnetization. This causes the hydrogen atoms to emit a weak radio signal which is amplified by the scanner. This signal can be manipulated by additional magnetic fields to build up enough information to reconstruct an image of the body.

Magnetic resonance spectroscopy is used to measure the levels of different metabolites in body tissues. The MR signal produces spectrum of difference resonances that correspond to different molecular arrangements of the isotope being "excited".

This signature is used to diagnose certain metabolic disorders, especially those affecting the brain, as well as to provide information on tumor metabolism.

The scanners used in medicine have a typical magnetic field strength of 0.2 to 3 teslas. Construction costs approximately US$ 1 million per tesla and maintenance an additional several hundred thousand dollars per year. Research using MRI scanners operating at ultra high field strength (up to 21.1 tesla) can produce images of the mouse brain with a resolution of 18 micrometres.

Nomenclature

Magnetic resonance imaging was developed from knowledge gained in the study of nuclear magnetic resonance. In its early years MRI was referred to as nuclear magnetic resonance imaging (NMRI), but the word nuclear has been associated with ionizing radiation exposure, which is not used in an MRI, so to prevent patients from making a negative association between MRI and ionizing radiation, the word has been almost universally removed. Scientists still use the term *NMRI* when discussing non-medical devices operating on the same principles.

One of the inventors of MRI, Paul Lauterbur, originally named the technique *zeugmatography*, a Greek term meaning "that which is used for joining". The term referred to the interaction between the static and the gradient magnetic fields necessary to create an image, but the nomenclature never caught on.

Brief Explanation for the Layperson

MRI works by making certain atoms in the body emit radiowaves by using a powerful magnet. When a person is in the scanner some of the hydrogen atoms, which are mainly found in water, align with this magnetic field. A radiowave at just the right frequency makes these atoms resonate and absorb energy.

The atoms then release this energy in the form of a very weak radiowave that can be measured by the scanner and amplified. Extra magnetic fields are used to constantly change the magnetic field to allow images of the body to be reconstructed.

These fields are created by gradient coils which make the familiar banging sounds of an MRI scan. Contrast agents may be injected to demonstrate blood vessels or inflammation in the tissues. Unlike CT scanning MRI uses no ionizing radiation and is generally a very safe procedure. Patients with some metal implants and cardiac pacemakers are prevented from having an MRI due to effect of the powerful magnetic field.

MRI is used to image every part of the body, but is particularly useful in neurological conditions, disorders of the muscles and joints, for evaluating tumors and showing abnormalities in the heart and blood vessels.

PRINCIPLE

Magnetism

Elementary subatomic particles such as protons have the quantum mechanical property of spin. Nuclei such as ^{1}H or ^{31}P, with an odd number of nucleons, always have a non–zero spin and therefore a magnetic moment. Some other isotopes such as ^{12}C have no unpaired neutrons or protons, and no net spin.

When these spins are placed in an external magnetic field they start to precess around the direction of that field. The magnetic field also creates two energy states that the protons can occupy which are separated by a quantum of energy. The thermal energy of the sample causes the molecules to tumble leaving only a very small excess of protons to cause magnetic polarization.

Resonance

The energy difference between the proton energy states corresponds to electromagnetic radiation at radio frequency wavelengths. Resonant absorption of energy by the protons due to an external oscillating magnetic field (radio wave) will occur at the Larmor frequency.

The net magnetization vector has two components. The longitudinal magnetization is due to an excess of protons in the lower energy state. This gives a net polarization parallel to the external field. The transverse magnetization is due to coherences forming between the two proton energy states. This gives a net polarization perpendicular to external field in the transverse plane. The recovery of longitudinal magnetization is called T1 relaxation and the loss of phase coherence in the transverse plane is called T2 relaxation.

When the radio frequency pulse is turned off, the transverse vector component produces an oscillating magnetic field which induces a small current in the receiver coil. This free induction decay (FID) lasts only a few milliseconds before the thermal equilibrium of the spins is restored. The actual signal that is measured by the scanner is formed by a refocusing gradient or radio wave to create a gradient or spin-echo.

Imaging

Slice selection is achieved by applying a magnetic gradient in addition to the external magnetic field during the radio frequency pulse. Only one plane within the object will have protons that are on–resonance and contribute to the signal.

A real image can be considered as being composed of a number of spatial frequencies at different orientations. A two–dimensional Fourier transformation of a real image will express these waves as a matrix of spatial frequencies known as k–space.

Low spatial frequencies are represented at the centre of k–space and high spatial frequencies at the periphery. Frequency and phase encoding are used to measure the amplitudes of a range of spatial frequencies within the object being imaged.

The frequency encoding gradient is applied during readout of the signal and is orthogonal to the slice selection gradient. During application of the gradient the frequency differences in the readout direction progressively change.

At the midpoint of the readout these differences are small and the low spatial frequencies in the image are sampled filling the centre of k-space. Higher spatial frequencies will be sampled towards the beginning and end of the readout filling the periphery of k-space.

Phase encoding is applied in the remaining orthogonal plane and uses the same principle of sampling the object for different spatial frequencies. However, it is applied for a brief period before the readout and the strength of the gradient is changed incrementally between each radio frequency pulse. For each phase encoding step a line of k–space is filled.

Magnet

The magnet is the largest and most expensive component of the scanner, and the remainder of the scanner is built around it. Just as important as the strength of the main magnet is its precision. The straightness of magnet lines within the centre or, as it is known as, the iso-centre of the magnet, needs to be nearly perfect. This is known as homogeneity. Fluctuations (non-homogeneities in the field strength) within the scan region should be less than three parts per million (3 ppm). Three types of magnet have been used:

- *Permanent magnet*: Conventional magnets made from ferromagnetic materials (e.g., steel) can be used to provide the static magnetic field. These are extremely bulky (the magnet can weigh in excess of 100 tonnes), but once installed require little costly maintenance. Permanent magnets can only achieve limited field strength (usually < 0.4 T) and have limited stability and precision. There are also potential safety issues, as the magnetic field cannot be removed in case of entrapment.
- *Resistive Electromagnet*: A solenoid wound from copper wire is an alternative to a permanent magnet. The advantages are low cost, but field strength is limited, and stability is poor. The electromagnet requires considerable electrical energy during operation which can make it expensive to operate. This design is essentially obsolete.
- *Superconducting Electromagnet*: When a niobium-titanium alloy is cooled by liquid helium at 4 K ("269 °C, "452 °F) it becomes superconducting where it loses all resistance to flow of electrical current. By building an electromagnet from superconducting wire, it is possible to develop extremely high field strengths, with very high stability. The construction of such magnets is extremely costly, and the cryogenic helium is expensive and difficult to handle. However, despite its cost, helium cooled superconducting magnets are the most common type found in MRI scanners today.

Most superconducting magnets have their coils of superconductive wire immersed in liquid helium, inside a vessel called a cryostat. Despite thermal insulation, ambient heat causes the helium to slowly boil off. Such magnets, therefore, require regular topping-up with helium.

Generally a cryocooler, also known as a coldhead is used to

recondense some helium vapour back into the liquid helium bath. Several manufacturers now offer 'cryogenless' scanners, where instead of being immersed in liquid helium the magnet wire is cooled directly by a cryocooler.

Magnets are available in a variety of shapes. However, permanent magnets are most frequently 'C' shaped, and superconducting magnets most frequently cylindrical. However, C-shaped superconducting magnets and box-shaped permanent magnets have also been used.

Magnetic field strength is an important factor determining image quality. Higher magnetic fields increase signal-to-noise ratio, permitting higher resolution or faster scanning.

However, higher field strengths require more costly magnets with higher maintenance costs, and have increased safety concerns. 1.0 - 1.5 T field strengths are a good compromise between cost and performance for general medical use. However, for certain specialist uses (e.g., brain imaging), field strengths up to 3.0 T may be desirable.

RADIO FREQUENCY SYSTEM

The radio frequency (RF) transmission system consists of a RF synthesizer, power amplifier and transmitting coil. This is usually built into the body of the scanner. The power of the transmitter is variable, but high-end scanners may have a peak output power of up to 35 kW, and be capable of sustaining average power of 1 kW.

The receiver consists of the coil, pre-amplifier and signal processing system. While it is possible to scan using the integrated coil for transmitting and receiving, if a small region is being imaged then better image quality is obtained by using a close-fitting smaller coil. A variety of coils are available which fit around parts of the body, e.g., the head, knee, wrist, or internally, e.g., the rectum.

A recent development in MRI technology has been the development of sophisticated multi-element phased array coils which are capable of acquiring multiple channels of data in parallel. This 'parallel imaging' technique uses unique acquisition schemes that allow for accelerated imaging, by replacing some of the spatial coding originating from the magnetic gradients with the spatial sensitivity of the different coil elements. However the increased acceleration also reduces the signal-to-noise ratio and can create residual artifacts in the image reconstruction. Two frequently used parallel acquisition and reconstruction schemes are SENSE and GRAPPA.

Gradients

Magnetic gradients are generated by three orthogonal coils, oriented in the x, y and z directions of the scanner. These are usually resistive electromagnets powered by sophisticated amplifiers which permit rapid and precise adjustments to their field strength and direction.

Typical gradient systems are capable of producing gradients from 20 mT/m to 100 mT/m (i.e. in a 1.5 T magnet, when a maximal z-axis gradient is applied the field strength may be 1.45 T at one end of a 1 m long bore, and 1.55 T at the other). It is the magnetic gradients that determine the plane of imaging - because the orthogonal gradients can be combined freely, any plane can be selected for imaging.

Scan speed is dependent on performance of the gradient system. Stronger gradients allow for faster imaging, or for higher resolution; similarly, gradients systems capable of faster switching can also permit faster scanning. However, gradient performance is limited by safety concerns over nerve stimulation.

In order to understand MRI contrast, it is important to have some understanding of the time constants involved in relaxation processes that establish equilibrium following RF excitation. As the high-energy nuclei relax and realign they emit energy at rates which are recorded to provide information about the material they are in.

The realignment of nuclear spins with the magnetic field is termed *longitudinal relaxation* and the time required for a certain percentage of the tissue's nuclei to realign is termed "Time 1" or T1, which is typically about 1 second at 1.5 Tesla main field strength. T2-weighted imaging relies upon local dephasing of spins following the application of the transverse energy pulse; the *transverse* relaxation time is termed "Time 2" or T2, typically $<$ 100 ms for tissue at 1.5 Tesla main field strength.

A subtle but important variant of the T2 technique is called T2 imaging. T2 imaging employs a spin echo technique, in which spins are refocused to compensate for local magnetic field inhomogeneities. T2 imaging is performed without refocusing. This sacrifices some image integrity (resolution) but provides additional sensitivity to relaxation processes that cause incoherence of transverse magnetization.

Applications of T2 imaging include functional MRI (fMRI) or evaluation of baseline vascular perfusion (e.g. cerebral blood flow (CBF)) and cerebral blood volume (CBV) using injected agents; in these cases, there is an inherent trade-off between image quality and detection sensitivity. Because T2-weighted sequences are sensitive to magnetic inhomogeneity (as can be caused by deposition of iron-containing blood-degradation products), such sequences are utilized to detect subtle areas of recent or chronic intra cranial hemorrhage ("Heme sequence").

Image contrast is created by using a selection of image acquisition parameters that weights signal by T1, T2 or T2, or no relaxation time ("proton-density images"). In the brain, T1-weighting causes the nerve connections of white matter to appear white, and the congregations of neurons of gray matter to appear gray, while cerebrospinal fluid appears dark.

The contrast of "white matter," "gray matter'" and "cerebrospinal fluid" is reversed using T2 or T2 imaging, whereas proton-weighted imaging provides little contrast in normal subjects. Additionally, functional information (CBF, CBV, blood oxygenation) can be encoded within T1, T2, or T2.

Diffusion weighted imaging (DWI) uses very fast scans with an additional series of gradients (diffusion gradients) rapidly turned on and off. Protons from water diffusing randomly within the brain, via Brownian motion, lose phase coherence and, thus signal during application of diffusion gradients.

In a brain with an acute infarction water diffusion is impaired, and signal loss on DWI sequences is less than in normal brain. DWI is the most sensitive method of detecting cerebral infarction (stroke) and works within 30 minutes of the ictus.

Contrast Enhancement

Both T1-weighted and T2-weighted images are acquired for most medical examinations; However they do not always adequately show the anatomy or pathology. The first option is to use a more sophisticated image acquisition technique such as fat suppression or chemical-shift imaging. The other is to administer a contrast agent to delineate areas of interest.

A contrast agent may be as simple as water, taken orally, for imaging the stomach and small bowel although substances with specific magnetic properties may be used. Most commonly, a paramagnetic contrast agent (usually a gadolinium compound) is given. Gadolinium-enhanced tissues and fluids appear extremely bright on T1-weighted images. This provides high sensitivity for detection of vascular tissues (e.g. tumors) and permits assessment of brain perfusion (e.g. in stroke). There have been concerns raised recently regarding the toxicity of gadolinium-based contrast agents and their impact on persons with impaired kidney function. Special actions may be taken, such as hemodialysis following a contrast MRI scan for renally-impaired patients.

More recently, superparamagnetic contrast agents (e.g. iron oxide nanoparticles) have become available. These agents appear very dark on T2-weighted images and may be used for liver imaging - normal liver tissue retains the agent, but abnormal areas (e.g. scars, tumors) do not. They can also be taken orally, to improve visualization of the gastrointestinal tract, and to prevent water in the gastrointestinal tract from obscuring other organs (e.g. pancreas).

Diamagnetic agents such as barium sulfate have been studied for potential use in the gastrointestinal tract, but are less frequently used.

MRI vs CT

A computed tomography (CT) scanner uses X-rays, a type of ionizing radiation, to acquire its images, making it a good tool for examining tissue composed of elements of a relatively higher atomic number than the tissue surrounding them, such as bone and calcifications (calcium based) within the body (carbon based flesh), or of structures (vessels, bowel). MRI, on the other hand, uses non-ionizing radio frequency (RF) signals to acquire its images and is best suited for non-calcified tissue.

CT may be enhanced by use of contrast agents containing elements of a higher atomic number than the surrounding flesh such as iodine or barium. Contrast agents for MRI are those which have paramagnetic properties. One example is gadolinium.

Both CT and MRI scanners can generate multiple two-dimensional cross-sections (slices) of tissue and three-dimensional reconstructions. Unlike CT, which uses only X-ray attenuation to generate image contrast, MRI has a long list of properties that may be used to generate image contrast. By variation of scanning parameters, tissue contrast can be altered and enhanced in various ways to detect different features.

MRI can generate cross-sectional images in any plane (including oblique planes). CT was limited to acquiring images in the axial (or near axial) plane in the past. The scans used to be called Computed *Axial* Tomography scans (CAT scans). However, the development of multi-detector CT scanners with near-isotropic resolution, allows the CT scanner to produce data that can be retrospectively reconstructed in any plane with minimal loss of image quality.

For purposes of tumor detection and identification, MRI is generally superior. However, CT usually is more widely available, faster, much less expensive, and may be less likely to require the person to be sedated or anesthetized.

MRI is also best suited for cases when a patient is to undergo the exam several times successively in the short term, because, unlike CT, it does not expose the patient to the hazards of ionizing radiation.

Economics of MRI

MRI equipment is expensive. New 1.5 Tesla scanners often cost between $1,000,000 USD and $1,500,000 USD. New 3.0 Tesla scanners often cost between $2,000,000 and $2,300,000 USD. Construction of MRI suites can cost $500,000 USD. For over a dozen years, MRI scanners have been significant sources of revenue for healthcare providers in the US. This is because of favorable reimbursement rates from insurers, both private and federal government programs. Insurance reimbursement has historically been provided in two components, technical for the actual performance of the MRI scan and professional for the radiologist's review of the images and/or data. In the US, the 2007 Deficit Reduction Act (DRA) significantly reduced reimbursement rates paid by federal insurance programs for the technical component of many scans, shifting the economic landscape. Many private insurers have followed suit.

Currently, in the US, there is increasing interest in reducing the costs associated with MRI services and simultaneously improving the ability to effectively and efficiently provide MRI examination services to larger numbers of patients with the same equipment.

The k-space Formalism

In 1983 Ljunggren and Tweig independently introduced the k-space formalism, a technique that proved invaluable in unifying different MR imaging techniques.

They showed that the demodulated MR signal $S(t)$ generated by freely precessing nuclear spins in the presence of a linear magnetic field gradient G equals the Fourier transform of the effective spin density P i.e.

$$S(t) = \tilde{\rho}\text{effective}(\vec{k}(t)) \equiv \int d^3x \rho(\vec{x}) \cdot e^{2\pi_2 \vec{k}(t) \cdot \vec{x}}$$

where:

$$\vec{k}(t) \equiv \int_0^t \vec{G}(t')dt'$$

In other words, as time progresses the signal traces out a trajectory in k-space with the velocity vector of the trajectory proportional to the vector of the applied magnetic field gradient.

By the term *effective spin density* we mean the true spin density $\rho(\vec{x})$ corrected for the effects of T_1 preparation, T_2 decay, dephasing due to field inhomogeneity, flow, diffusion, etc. and any other phenomena that affect that amount of transverse magnetization available to induce signal in the RF probe.

From the basic k-space formula, it follows immediately that we reconstruct an image $I(\vec{x})$ simply by taking the inverse Fourier transform of the sampled data viz.

$$I(\vec{x}) = \int d^3k S(\vec{k}(t)) \cdot e^{-2\pi_2 \vec{k}(t) \cdot \vec{x}}$$

Using the k-space formalism, a number of seemingly complex ideas become simple. For example, it becomes very easy to understand the role of phase encoding (the so-called spin-warp method). In a standard spin echo or gradient echo scan, where the readout (or view) gradient is constant (e.g. G_x), a single line of k-space is scanned per RF excitation. When the phase encoding gradient is zero, the line scanned is the k_x axis. When a non-zero phase-encoding pulse is added in between the RF excitation and the commencement of the readout gradient, this line moves up or down in k-space i.e. we scan the line k_y=constant.

The k-space formalism also makes it very easy to compare different scanning techniques. In single-shot EPI, all of k-space is scanned in a single shot, following either a sinusoidal or zig-zag trajectory. Since alternating lines of k-space are scanned in opposite directions, this must be taken into account in the reconstruction.

Multi-shot EPI and fast spin echo techniques acquire only part of k-space per excitation. In each shot, a different interleaved segment is acquired, and the shots are repeated until k-space is sufficiently well-covered. Since the data at the centre of k-space represent lower spatial frequencies than the data at the edges of k-space, the T_E value for the centre of k-space determines the image's T_2 contrast.

The importance of the centre of k-space in determining image contrast can be exploited in more advanced imaging techniques. One such technique is spiral acquisition - a rotating magnetic field gradient is applied, causing the trajectory in k-space to trace out spiral out from the centre to the edge.

Due to T_2 and T_2 decay the signal is greatest at the start of the acquisition, hence acquiring the centre of k-space first improves contrast to noise ratio (CNR) when compared to conventional zig-zag acquisitions, especially in the presence of rapid movement.

Since $\vec{x}$ and $\vec{k}$ are conjugate variables (with respect to the Fourier transform) we can use the Nyquist theorem to show that the step in k-space determines the field of view of the image (maximum frequency that is correctly sampled) and the maximum value of k sampled determines the resolution i.e.

$$\mathrm{FOV} \propto \frac{1}{\Delta k} \quad \mathrm{Resolution} \propto |k_{max}|$$

(these relationships apply to each axis [X, Y, and Z] independently).

Chapter 14

Application of MRI

In clinical practice, MRI is used to distinguish pathologic tissue (such as a brain tumor) from normal tissue. One advantage of an MRI scan is that it is thought to be harmless to the patient. It uses strong magnetic fields and non-ionizing radiation in the radio frequency range. Compare this to CT scans and traditional X-rays which involve doses of ionizing radiation and may increase the risk of malignancy, especially in a fetus.

While CT provides good spatial resolution (the ability to distinguish two structures an arbitrarily small distance from each other as separate), MRI provides comparable resolution with far better contrast resolution (the ability to distinguish the differences between two arbitrarily similar but not identical tissues). The basis of this ability is the complex library of *pulse sequences* that the modern medical MRI scanner includes, each of which is optimized to provide *image contrast* based on the chemical sensitivity of MRI.

For example, with particular values of the *echo time* (TE) and the *repetition time* (TR), which are basic parameters of image acquisition, a sequence will take on the property of T2-weighting. On a T2-weighted scan, fat-, water- and fluid-containing tissues are bright (most modern T2 sequences are actually *fast* T2 sequences).

Damaged tissue tends to develop edema, which makes a T2-weighted sequence sensitive for pathology, and generally able to distinguish pathologic tissue from normal tissue.

With the addition of an additional radio frequency pulse and additional manipulation of the magnetic gradients, a T2-weighted sequence can be converted to a FLAIR sequence, in which free water is now dark, but edematous tissues remain bright. This sequence in particular is currently the most sensitive way to evaluate the brain for demyelinating diseases, such as multiple sclerosis.

The typical MRI examination consists of 5-20 sequences, each of which are chosen to provide a particular type of information about the subject tissues. This information is then synthesized by the interpreting physician.

Specialized MRI scans

Diffusion MRI

Diffusion MRI measures the diffusion of water molecules in biological tissues. In an isotropic medium (inside a glass of water for example) water molecules naturally move randomly according to Brownian motion. In biological tissues however, the diffusion may be anisotropic. For example a molecule inside the axon of a neuron has a low probability of crossing the myelin membrane.

Therefore the molecule will move principally along the axis of the neural fibre. If we know that molecules in a particular voxel diffuse principally in one direction we can make the assumption that the majority of the fibers in this area are going parallel to that direction.

The recent development of diffusion tensor imaging (DTI) enables diffusion to be measured in multiple directions and the fractional anisotropy in each direction to be calculated for each voxel. This enables researchers to make brain maps of fibre directions to examine the connectivity of different regions in the brain (using tractography) or to examine areas of neural degeneration and demyelination in diseases like Multiple Sclerosis.

Another application of diffusion MRI is diffusion-weighted imaging (DWI). Following an ischemic stroke, DWI is highly sensitive to the changes occurring in the lesion. It is speculated that increases in restriction (barriers) to water diffusion, as a result of cytotoxic edema (cellular swelling), is responsible for the increase in signal on a DWI scan.

Other theories, including acute changes in cellular permeability and loss of energy-dependent (ATP) cytoplasmic streaming, have been proposed to explain the phenomena. The DWI enhancement appears within 5-10 minutes of the onset of stroke symptoms (as compared with computed tomography, which often does not detect changes of acute infarct for up to 4-6 hours) and remains for up to two weeks.

CT, due to its insensitivity to acute ischemia, is typically employed to rule out hemorrhagic stroke, which would entirely prevent the use of tissue plasminogen activator (tPA). Further, coupled with scans sensitized to cerebral perfusion, researchers can highlight regions of "perfusion/ diffusion mismatch" that may indicate regions capable of salvage by reperfusion therapy.

Finally, it has been proposed that diffusion MRI may be able to detect minute changes in extracellular water diffusion and therefore could be used as a tool for fMRI. The nerve cell body enlarges when it conducts an action potential, hence restricting extracellular water molecules from diffusing naturally. Although this process works in theory, evidence is only moderately convincing. Like many other specialized applications, this technique is usually coupled with a fast image acquisition sequence, such as echo planar imaging sequence.

Magnetic Resonance Angiography

Magnetic resonance angiography (MRA) is used to generate pictures of the arteries in order to evaluate them for stenosis (abnormal narrowing) or aneurysms (vessel wall dilatations, at risk of rupture). MRA is often used to evaluate the arteries of the neck and brain, the thoracic and abdominal aorta, the renal arteries, and the legs (called a "run-off").

A variety of techniques can be used to generate the pictures, such as administration of a paramagnetic contrast agent (gadolinium) or using a technique known as "flow-related enhancement" (e.g. 2D and 3D time-of-flight sequences), where most of the signal on an image is due to blood which has recently moved into that plane.

Magnetic resonance venography (MRV) is a similar procedure that is used to image veins. In this method the tissue is now excited inferiorly while signal is gathered in the plane immediately superior to the excitation plane, and thus imaging the venous blood which has recently moved from the excited plane.

Magnetic Resonance Spectroscopy

In vivo ('in the living organism') magnetic resonance spectroscopy (MRS), also known as MRSI (MRS imaging) and volume selective NMR spectroscopy, is a technique which combines the spatially-addressable nature of MRI with the spectroscopically-rich information obtainable from NMR.

That is to say, MRI allows one to study a particular region within an organism or sample, but gives relatively little information about the chemical or physical nature of that region (its chief value is in being able to distinguish the properties of that region, how much fat or water is present, relative to those of surrounding regions).

MR spectroscopy, however, provides a wealth of information about other biological chemicals ('metabolites') within that region, as would an NMR spectrum of that region.

Functional MRI

Functional MRI (fMRI) measures signal changes in the brain that are due to changing neural activity. The brain is scanned at low resolution but at a rapid rate (typically once every 2-3 seconds). Increases in neural activity cause changes in the MR signal via T2 changes; this mechanism is referred to as the BOLD (blood-oxygen-level dependent) effect.

Increased neural activity causes an increased demand for oxygen, and the vascular system actually overcompensates for this, increasing the amount of oxygenated hemoglobin relative to deoxygenated hemoglobin.

Because deoxygenated hemoglobin attenuates the MR signal, the vascular response leads to a signal increase that is related to the neural activity. The precise nature of the relationship between neural activity and the BOLD signal is a subject of current research. The BOLD effect also allows for the generation of high resolution 3D maps of the venous vasculature within neural tissue.

While BOLD signal is the most common method employed for neuroscience studies in human subjects, the flexible nature of MR imaging provides means to sensitize the signal to other aspects of the blood supply. Alternative techniques employ arterial spin labeling (ASL) or weight the MRI signal by cerebral blood flow (CBF) and cerebral blood volume (CBV).

The CBV method requires injection of a class of MRI contrast agents that are now in human clinical trials. Because this method has been shown to be far more sensitive than the BOLD technique in preclinical studies, it may potentially expand the role of fMRI in clinical applications. The CBF method provides more quantitative information than the BOLD signal, albeit at a significant loss of detection sensitivity.

Interventional MRI

The lack of harmful effects on the patient and the operator make MRI well-suited for "interventional radiology", where the images produced by a MRI scanner are used to guide minimally-invasive procedures.

Radiation Therapy Simulation

Because of MRI's superior imaging of soft tissues, it is now being utilized to specifically locate tumors within the body in preparation for radiation therapy treatments. For therapy simulation, a patient is placed in specific, reproducible, body position and scanned.

The MRI system then computes the precise location, shape and orientation of the tumor mass, correcting for any spatial distortion inherent in the system. The patient is then marked or tattooed with points which, when combined with the specific body position, will permit precise triangulation for radiation therapy.

Current Density Imaging

Current density imaging (CDI) endeavors to use the phase information from images to reconstruct current densities within a subject. Current density imaging works because electrical currents generate magnetic fields, which in turn affect the phase of the magnetic dipoles during an imaging sequence. To date no successful CDI has been performed using biological currents, but several studies have been published which involve applied currents through a pair of electrodes.

Magnetic Resonance Guided Focused Ultrasound

In MRgFUS therapy, ultrasound beams are focused on a tissue - guided and controlled using MR thermal imaging - and due to the significant energy deposition at the focus, temperature within the tissue rises to more than 65°C, completely destroying it. This technology can achieve precise "ablation" of diseased tissue.

MR imaging provides a three-dimensional view of the target tissue, allowing for precise focusing of ultrasound energy. The MR imaging provides quantitative, real-time, thermal images of the treated area. This allows the physician to ensure that the temperature generated during each cycle of ultrasound energy is sufficient to cause thermal ablation within the desired tissue and if not, to adapt the parameters to ensure effective treatment.

Multinuclear Imaging

Hydrogen is the most frequently imaged nucleus in MRI because it is present in biological tissues in great abundance. However, any nucleus which has a net nuclear spin could potentially be imaged with MRI. Such nuclei include helium-3, carbon-13, fluorine-19, oxygen-17, sodium-23, phosphorus-31 and xenon-129. ^{23}Na and ^{31}P are naturally abundant in the body, so can be imaged directly.

Gaseous isotopes such as ^{3}He or ^{129}Xe must be hyperpolarized and then inhaled as their nuclear density is too low to yield a useful signal under normal conditions. ^{17}O, ^{13}C and ^{19}F can be administered in sufficient quantities in liquid form (e.g. ^{17}O-water, ^{13}C-glucose solutions or perfluorocarbons) that hyperpolarization is not a necessity.

Multinuclear imaging is primarily a research technique at present. However, potential applications include functional imaging and imaging of organs poorly seen on ^{1}H MRI (e.g. lungs and bones) or as alternative contrast agents. Inhaled hyperpolarized ^{3}He can be used to image the distribution of air spaces within the lungs. Injectable solutions containing ^{13}C or stabilized bubbles of hyperpolarized ^{129}Xe have been studied as contrast agents for angiography and perfusion imaging. ^{31}P can potentially provide information on bone density and structure, as well as functional imaging of the brain.

Experimental MRI Techniques

Currently there is active research in several new MRI technologies like magnetization transfer MRI (MT-MRI), diffusion tensor MRI (DT-MRI), and proton MR spectroscopy, plus recent research in to Dendrimer-enhanced MRI as a diagnostic and prognostic biomarker of sepsis-induced acute renal failure.

SAFETY

Implants and foreign bodies: Pacemakers are generally considered an absolute contraindication towards MRI scanning, though highly specialized protocols have been developed to permit scanning of select pacing devices. Several cases of arrhythmia or death have been reported in patients with pacemakers who have undergone MRI scanning without appropriate precautions. Other electronic implants have varying contraindications, depending upon scanner technology, implant properties, scanning protocols and anatomy being imaged.

Though pacemakers receive significant attention, it should also be noted that many other forms of medical or biostimulation implants may be contraindicated for MRI scans. These may include Vagus nerve stimulators, implantable cardioverter-defibrillators (ICD), loop recorders, insulin pumps, cochlear implants, deep brain stimulators and many others.

Medical device patients should always present complete information (manufacturer, model, serial number and date of implantation) about all implants to both the referring physician and to the radiologist or technologist before entering the room for the MRI scan.

While these implants pose a current problem, scientist and manufacturers are working on improved designs which will further minimize the risks that MRI scans pose to medical device operations. One such development in the works is a nano coating for implants intended to screen them from the radio frequency waves, helping to make MRI exams available to patients currently prohibited from receiving them. The current article for this is from New Scientist.

Ferromagnetic foreign bodies (e.g. shell fragments), or metallic implants (e.g. surgical prostheses, aneurysm clips) are also potential risks, and safety aspects need to be considered on an individual basis. Interaction of the magnetic and radio frequency fields with such objects can lead to: trauma due to movement of the object in the magnetic field, thermal injury from radio-frequency induction heating of the object, or failure of an implanted device.

These issues are especially problematic when dealing with the eye. Most MRI centers require an orbital x-ray be performed on anyone who suspects they may have small metal fragments in their eyes, perhaps from a previous accident, something not uncommon in metalworking.

Because of its non-ferromagnetic nature and poor electrical

conductivity, titanium and its alloys are useful for long term implants and surgical instruments intended for use in image-guided surgery.

In particular, not only is titanium safe from movement from the magnetic field, but artifacts around the implant are less frequent and less severe than with more ferromagnetic materials e.g. stainless steel. Artifacts from metal frequently appear as regions of empty space around the implant - frequently called 'black-hole artifact' e.g. a 3mm titanium alloy coronary stent may appear as a 5mm diameter region of empty space on MRI, whereas around a stainless steel stent, the artifact may extend for 10-20 mm or more.

Projectile or Missile Effect

As a result of the very high strength of the magnetic field needed to produce scans (frequently up to 60,000 times the earth's own magnetic field effects), there are several incidental safety issues addressed in MRI facilities. Missile-effect accidents, where ferromagnetic objects are attracted to the centre of the magnet, have resulted in injury and death. A video simulation of a fatal projectile effect accident illustrates the extreme power that contemporary MRI equipment can exert on ferromagnetic objects.

In order to help reduce the risks of projectile accidents, ferrous objects and devices are typically prohibited in proximity to the MRI scanner, with non ferro-magnetic versions of many tools and devices typically retained by the scanning facility. Patients undergoing MRI examinations are required to remove all metallic objects, often by changing into a gown or 'scrubs'.

The magnetic field and the associated risk of missile-effect accidents remains a permanent hazard — as superconductive MRI magnets retain their magnetic field, even in the event of a power outage.

Radio Frequency Energy

A powerful radio transmitter is needed for excitation of proton spins. This can heat the body significantly, with the risk of hyperthermia in patients, particularly the obese or patients with thermoregulation disorders. Several countries have issued restrictions on the maximum specific absorption rate that a scanner may produce.

Peripheral Nerve Stimulation

The rapid switching (on and off) of the magnetic field gradients needed for imaging is capable of causing nerve stimulation. Volunteers report a twitching sensation when exposed to rapidly switched fields, particularly in their extremities. The reason the peripheral nerves are stimulated is that the changing field increases with distance from the centre of the gradient coils (which more or less coincides with the centre of the magnet).

Note however that when imaging the head, the heart is far off-centre and induction of even a tiny current into the heart must be avoided at all costs. Although PNR was not a problem for the slow, weak gradients used in the early days of MRI, the strong, rapidly-switched gradients used in techniques such as EPI, fMRI, diffusion MRI, etc. are indeed capable of inducing PNR.

American and European regulatory agencies insist that manufacturers stay below specified dB/dt limits (dB/dt is the change in field per unit time) or else prove (via clinical studies) that no PNR is induced for any imaging sequence. As a result of dB/dt limitation software and/or hardware, commercial MRI systems cannot use the full rated power of their gradient amplifiers.

Acoustic Noise

Loud noises and vibrations are produced by forces resulting from rapidly switched magnetic gradients interacting with the main magnetic field, in turn causing minute expansions and contractions of the coil itself. This is most marked with high-field machines and rapid-imaging techniques in which sound intensity can reach 130 dB (equivalent to a jet engine at take-off).

Appropriate use of ear protection is essential for anyone inside the MRI scanner room during the examination.

Cryogens

As described above in 'Scanner Construction And Operation', many MRI scanners rely on cryogenic liquids to enable superconducting capabilities of the electromagnetic coils within. Though the cryogenic liquids most frequently used are non-toxic, their physical properties present specific hazards.

An emergency shut-down of a superconducting electromagnet, an operation known as "quenching", involves the rapid boiling of liquid helium from the device. If the rapidly expanding helium cannot be dissipated though an external vent, sometimes referred to as 'quench pipe', it may be released into the scanner room where it may cause displacement of the oxygen and present a risk of asphyxiation.

Liquid helium, the most commonly used cryogen in MRI, undergoes near explosive expansion as it changes from liquid to a gaseous state. Rooms built in support of superconducting MRI equipment should be equipped with pressure relief mechanisms and an exhaust fan, in addition to the required quench pipe. Since a quench results in rapid loss of all cryogens in the magnet, recommissioning the magnet is extremely expensive and time-consuming.

Spontaneous quenches are uncommon, but can occur at any time. Quenches may also be triggered by equipment malfunction, improper cryogen fill technique, contaminates inside the cryostat, or extreme magnetic or vibrational disturbances.

Contrast agents

The most frequently used intravenous contrast agents are based on chelates of gadolinium. In general, these agents have proved safer than the iodinated contrast agents used in X-ray radiography or CT. Anaphylactoid reactions are rare occurring in approx 0.03-0.1%.

Of particular interest is the lower incidence of nephrotoxicity, compared with iodinated agents, when given at usual doses—this has made contrast-enhanced MRI scanning an option for patients with renal impairment, who would otherwise not be able to undergo contrast-enhanced CT.

Although gadolinium agents have proved useful for patients with renal impairment, in patients with severe renal failure requiring dialysis there is a risk of a rare but serious illness, nephrogenic systemic fibrosis, that may be linked to the use of certain gadolinium-containing agents: the most frequently linked is gadodiamide, but other agents have been linked too.

Although a causal link has not been definitively established, current guidelines in the United States are that dialysis patients should only receive gadolinium agents where essential, and that dialysis should be performed as soon as possible after the scan is complete, in order to remove the agent from the body promptly. In Europe where more gadolinium-containing agents are available, a classification of agents according to potential risks has been released.

Pregnancy

No harmful effects of MRI on the fetus have been demonstrated. In particular, MRI avoids the use of ionizing radiation, to which the fetus is particularly sensitive. However, as a precaution, current guidelines recommend that pregnant women undergo MRI only when essential. This is particularly the case during the first trimester of pregnancy, as organogenesis takes place during this period.

The concerns in pregnancy are the same as for MRI in general, but the fetus may be more sensitive to the effects—particularly to heating and to noise. However, one additional concern is the use of contrast agents; gadolinium compounds are known to cross the placenta and enter the fetal bloodstream, and it is recommended that their use be avoided.

Despite these concerns, MRI is rapidly growing in importance as a way of diagnosing and monitoring congenital defects of the fetus because it can provide more diagnostic information than ultrasound and it lacks the ionizing radiation of CT.

MRI without contrast is the imaging mode of choice for pre-surgical, in-utero diagnosis and evaluation of fetal tumors, primarily teratomas, facilitating open fetal surgery, other fetal interventions, and planning for procedures (such as the EXIT procedure) to safely deliver and treat babies whose defects would otherwise be fatal.

Claustrophobia and Discomfort

Due to the construction of MRI scanners, they are potentially unpleasant in which to lie. The part of the body being imaged needs to lie at the centre of the magnet (which is often a long, narrow tube). Because scan times may be long (perhaps one hour), people with even mild claustrophobia are often unable to tolerate an MRI scan without management.

Management may include:

- Advance preparation
  - Visiting the scanner to see the room and practice lying on the table
  - Visualization techniques
  - Chemical sedation
  - General anesthesia
- Coping while inside the scanner
  - Holding a "panic button"
  - listening to music on headphones or watching a movie with a Head-mounted display while in the machine
- Modified scanner designs
  - Upright MRIs (made exclusively by FONAR)
  - Open-bore design scanners.

Though open MRIs have increased in popularity as of late, they produce inferior scan quality because they operate at lower magnetic fields than closed MRIs.

For babies and children, chemical sedation or general anesthesia is the norm. These MRI subjects are too young to be instructed to hold still during the scanning session. Obese patients and pregnant women may find the MRI machine to be a tight fit, and even some claustrophobics may find the experience intolerable without sedation. Pregnant women may also have difficulty lying on their backs for an hour or more without moving. The noise associated with the operation of an MRI scanner (especially the audible noise associated with the gradient pulses applied to the subject) can also exacerbate the discomfort associated with the procedure.

The European Physical Agents Directive

The European Physical Agents (Electromagnetic Fields) Directive is European legislation that has been adopted in European legislature. By 2008 each individual state within the European Union must include this directive in its own law.

The directive applies to occupational exposure to electromagnetic fields (not medical exposure) and was intended to limit workers' acute exposure to strong electromagnetic fields, as may be found near electricity substations, radio or television transmitters or industrial equipment.

However, the regulations impact significantly on MRI, with separate sections of the regulations limiting exposure to static magnetic fields, changing magnetic fields and radio frequency energy. Field strength limits are given which may not be exceeded for any period of time. An employer may commit a criminal offense by allowing a worker to exceed an exposure limit if that is how the Directive is implemented in a particular Member State.

The Directive is based on the international consensus of established effects of exposure to electromagnetic fields, and in particular the advice of the European Commissions's advisor, the International Commission on Non-Ionizing Radiation Protection (ICNIRP).

The aims of the Directive, and the ICNIRP guidelines upon which it is based, are to prevent exposure to potentially harmful fields. The actual limits in the Directive are very similar to the limits advised by the Institute of Electrical and Electronics Engineers, with the exception of the frequencies produced by the gradient coils, where the IEEE limits are significantly higher.

Many Member States of the EU already have either specific EMF regulations or (as in the UK) a general requirement under workplace health and safety legislation to protect workers against electromagnetic fields. In almost all cases the existing regulations are aligned with the ICNIRP limits so that the Directive should, in theory, have little impact on any employer already meeting their legal responsibilities.

The introduction of the Directive has brought to light an existing potential issue with occupational exposures to MRI fields. There are at present very few data on the number or types of MRI practice that might lead to exposures in excess of the levels of the Directive. There is a justifiable concern amongst MRI practitioners that if the Directive were to be enforced more vigorously than existing legislation, the use of MRI might be restricted, or working practices of MRI personnel might have to change.

In the initial draft a limit of static field strength to 2 T was given. This has since been removed from the regulations, and whilst it is unlikely to be restored as it was without a strong justification, some restriction on static fields may be reintroduced after the matter has been considered more fully by ICNIRP.

The effect of such a limit might be to restrict the installation, operation and maintenance of MRI scanners with magnets of 2 T and stronger. As the increase in field strength has been instrumental in developing higher resolution and higher performance scanners, this would be a significant step back. This is why it is unlikely to happen without strong justification.

Individual government agencies and the European Commission have now formed a working group to examine the implications on MRI and to try to address the issue of occupational exposures to electromagnetic fields from MRI.

Chapter 15

Uses of Radioactivity

Radiation is all around us. It comes from the Earth and from outer space. Many forms of radiation are invisible — we can't feel it, taste it, or smell it. Yet, it can be detected and measured when present. We measure ionizing radiation in units called millirems. But what is radiation? Radioactive materials are composed of atoms that are unstable. An unstable atom gives off its excess energy until it becomes stable. The energy emitted is radiation. We can classify radiation as being either natural or man-made.

The Earth is surrounded by radiation. Every day, for example, we are exposed to radon, a radioactive gas from uranium found in soil dispersed in the air; from radioactive potassium in our food and water; from uranium, radium, and thorium in the Earth's crust; and from cosmic rays and the sun.

These types of radiation are called natural or background radiation. We are exposed to an average of 300 millirems of natural radiation each year (a millirem is a unit of measure for exposure to radiation). This amounts to natural radiation accounting for nearly 85 per cent of our total annual exposure. Where does the remaining 15 per cent come from? Man-made sources.

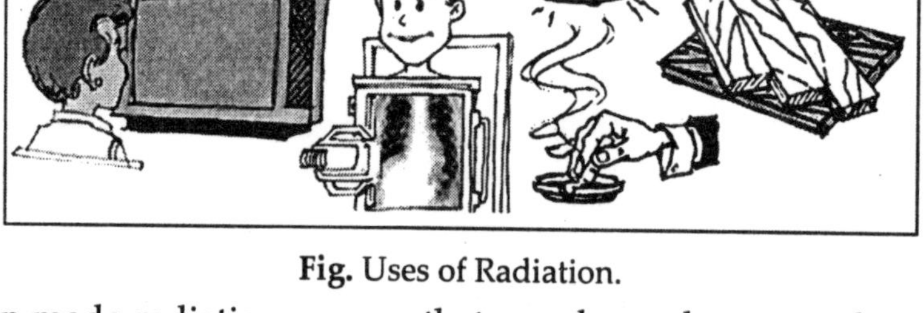

Fig. Uses of Radiation.

Man-made radiation sources that people can be exposed to include tobacco, television, medical x-rays, smoke detectors, lantern mantles, nuclear medicine, and building materials.

Generally, when we think of exposure to radiation, we need to look at radioactive atoms produced in nuclear reactors and described as being unstable. They are unstable because they undergo a disintegrating process called decaying. During this process, unstable atoms becomes stable, throwing off (emitting) radiation in the form of rays and/or particles.

How fast a radioactive atom decays into a stable atom depends on the atom itself. For example, the range in the rate of decay among isotopes goes from fractions of seconds to several billion years (e.g., uranium).

Let's take a look at uranium-238 to illustrate the decay chain.

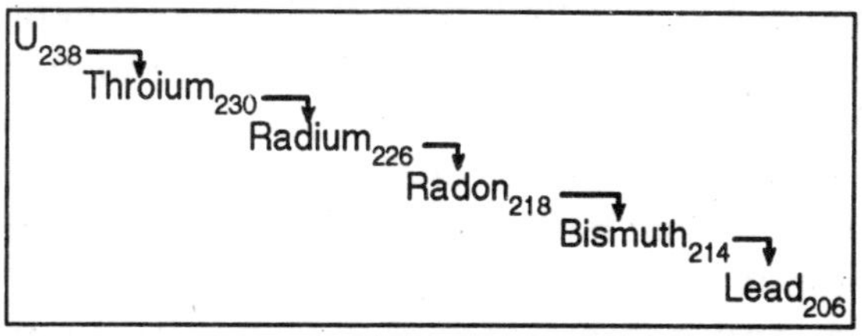

As U-238 decays it changes into thorium-230, which changes into radium-226, which changes into radon-218, which changes into bismuth-214, and finally into lead-206 (a stable element).

One peculiar thing about radioactive atoms is that no one knows exactly when the element will decay and give off radiation. There is, however, a pattern relating to how long it takes for an isotope to lose half of its radioactivity. The pattern is called half-life. If an atom, for example, has a half-life of 10 years, half of its atoms will decay in 10 years. Then in another 10 years half of that amount will decay and so on. While there are several different forms of radiation, we're going to concentrate on just three that result from the decay of radioactive isotopes: alpha, beta, and gamma.

Beta particles are high energy electrons. Both alpha and beta particles are emitted from unstable isotopes. The alpha particle, consisting of two protons and two neutrons, is relatively large compared to beta particles. Gamma rays have no mass.

Because of its size and electrical charge (+2), the alpha particle has a relatively slow speed and low penetrating distance (one to two inches in air). Alpha particles are easily stopped by a thin sheet of paper or the body's outer layer of skin. Since they do not penetrate the outer (dead) layer of skin, they present little or no hazard when they are external to the body.

However, alpha particles are considered internal hazards, because when they come into contact with live tissue they cause a large number of ionizations to occur in small areas, thus causing damage to tissues and cells.

Beta radiation, while faster and lighter than alpha radiation, can travel through about 10 feet of air and penetrate very thin layers of materials such as aluminum foil. However, while clothing will stop most beta particles, they can penetrate the live layers of skin tissue. Therefore, beta radiation is considered to be both an internal and external (to skin only) hazard. Thin layers of metals and plastics can be used to shield individuals from beta radiation.

Gamma radiation, high energy light, is a little different. It is a type of electromagnetic wave, just like radio waves, light waves, and x-rays. Gamma radiation is a very strong type of electromagnetic wave. It is has no weight and travels at the speed of light. This is much faster than alpha and beta radiation.

Because of their penetrating capability, gamma rays are considered both internal and external hazards. Thick walls of cement, lead, or steel are needed to stop it.

Alpha, beta, and gamma radiations are also known as ionizing radiation. Ionizing radiation is especially harmful because it can change the chemical makeup of many things, including the delicate chemistry of the human body and other living organisms. For this reason it is a good idea to avoid unnecessary exposure to all ionizing radiation.

The problem is simply this: large amounts of radiation — far above the levels encountered in daily life — can produce cancer and genetic defects in living organisms. Radiation causes damage and alters the body's normal cells and normal cell function. This breakdown in normal cell function may result in an uncontrolled growth of cells, hence the potential for malignant/cancerous tumors.

Whether the source of radiation is natural or man-made, whether it is a small dose of radiation or a large dose, there will be some biological effects. A diagram can show the biological effect of ionizing radiation.

Radiation causes ionizations of atoms that will affect molecules that may affect cells that may effect tissues that may affect organs that may affect the whole body.

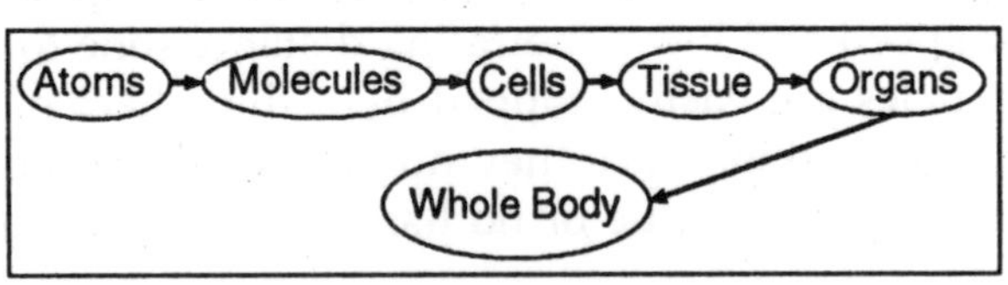

Fig. The Effect of Radiation on Living Cells.

Although we tend to think of biological effects in terms of the effect of radiation on living cells, in actuality, ionizing radiation, by definition, interacts only with atoms by a process called ionization. Thus, all biological damage effects begin when radiation interacts with atoms forming the cells in the human body. As a result, radiation effects on humans proceed from the lowest to the highest level as noted on the board. Although scientists have only known about radiation since the 1890s, they have developed a wide variety of uses for this natural force. Today, to benefit mankind, radiation is used in science, medicine, and industry, as well as for generating electricity. Radiation has useful applications in such areas as agriculture, medicine, space exploration, architect/engineering, industry/manufacturing, government, geology (including mining), ecology, and education.

Radiation is used by doctors to diagnose illness and helps archaeologists find the age of ancient artifacts. Electricity produced by nuclear fission — splitting the atom — is one of its greatest uses. A reliable source of electricity is needed to give us light, help to groom and feed us, and to keep our homes and businesses running. Let me give you some specific examples of how the radiation has been used to -

- Diagnose and treat illnesses.
- Kill bacteria and preserve food without chemicals and refrigeration.
- Process sludge for fertilizer and soil conditioner.
- Locate underground natural resources and tell a dry hole from a gusher.
- Make smoke detectors, nonstick fry pans, and ice cream.
- Grow stronger crops.
- Power satellites and provide future electrical needs for space laboratories with people on board.
- Design instruments, techniques, and equipment; measure air pollution.
- Prove the age of works of art and assist in determining their authenticity.

RADIATION IN SCIENCE

Radiation is used in science in many ways. Just as doctors can label substances inside people's bodies, scientists can label substances that pass through plants, animals, or our world. This allows us to study such things as the paths that different types of air and water pollution take through the environment.

It has also helped us learn more about a wide variety of things, such as what types of soil different plants need to grow, the size of newly discovered oil fields, and the track of ocean currents.

Scientists use radioactive substances to find the age of ancient objects by a process called carbon dating. For example, in the upper levels of our atmosphere, cosmic rays hit nitrogen atoms and form a naturally radioactive isotope called carbon-14. Carbon is found in all living things, and a small percentage of this carbon is carbon-14. When a plant or animal dies, it no longer takes in new carbon and the carbon-14 it contains begins the process of radioactive decay.

However, new isotopes of carbon-14 continue to be formed in our atmosphere, and after a few years the per cent of radioactivity in an old object is less than it is in a newer one. By measuring this difference, scientists are able to determine how old certain objects are. The measuring process is called carbon dating.

RADIATION USED TO SOLVE CRIMES

How is radiation used to solve crimes: I am sure that you already know that detectives often search the scene of a crime for traces of paint, glass, hair, gunpowder, or blood. But you may not know that after such evidence is collected, it is often exposed to radiation and then analysed to find out its exact makeup.

If material is exposed to streams of neutrons, some of the neutrons can be absorbed into the nucleus of the exposed material. This makes these materials slightly radioactive because they are unstable and decay with time. Scientists are then able to read the exact chemical signatures of these substances. This laboratory process, called activation analysis, is precise enough to tell if a single hair found at a crime scene came from a certain person.

Activation analysis is also used to find out the chemical makeup of materials when scientists only have small samples, as well as to prove that older works of art are not made of modern materials.

RADIATION IN INDUSTRY

We could talk all day about the many and varied uses of radiation in industry and not complete the list. To make a long story short, we'll just concentrate on a few. Exposure to some types of radiation (for example, x-rays) can kill germs without harming the items that are being disinfected or making them radioactive.

For example, when treated with radiation, foods take much longer to spoil, and medical equipment such as bandages, hypodermic syringes, and surgical instruments don't have to be exposed to toxic chemicals or extreme heat to be sterilized. Although we now use chlorine, a toxic and difficult-to-handle chemical, we may use radiation in the future to disinfect our drinking water and even kill all the germs in our sewage. Ultraviolet light already is used to disinfect drinking water in some homes.

The agricultural industry makes use of radiation to improve food production. Plant seeds, for example, have been exposed to radiation to bring about new and better types of plants. Besides making plants stronger, radiation can also be used to control insect populations, thereby decreasing the use of pesticides.

Engineers use radioactive substances to measure the thickness of materials and an x-ray process called radiography to find hard to detect defects in many types of metals and machines. Radiography is also used to check such things as the flow of oil in sealed engines and the rate and way various materials wear out. And we've already talked about the use of the radioactive element uranium, which is used as a fuel to make electricity for our cities, farms, towns, factories, etc.

In outer space, radioactive materials are also used to power space craft. Such materials have also been used to supply electricity to satellites sent on missions to the outermost regions of our solar system.

Radiation has been used to help clean up toxic pollutants, such as exhaust gases from coal-fired power stations and industry. Sulphur dioxides and nitrogen oxides, for example, can be removed by electron beam radiation.

As you can see, radiation and radioactive materials have played and will continue to play a very significant role in our lives. Let's sum up this discussion with a walk through the life of a typical family for one day and learn about some of the uses of radiation.

Dad gets up in the morning and puts on a clean shirt. His polyester-cotton blend shirt is made from chemically treated fabric that has been irradiated (treated with radiation) before being exposed to a soil-releasing agent. The radiation makes the chemicals bind to the fabric, keeping his shirt fresh and pressed all day. The shirt is not radioactive.

In the kitchen, Jenny is frying an egg. That nonstick pan she is using has been treated with gamma rays, and the thickness of the eggshell was measured by a gauge containing radioactive material before going into the egg carton. Thin, breakable eggs were screened out. The turkey Mom is taking out of the refrigerator for tonight's dinner was covered with irradiated polyethylene shrink wrap. Once polyethylene has been irradiated, it can be heated above its usual melting point and wrapped around the turkey to provide an airtight cover.

As Dad drives to work, he passes reflective signs that have been treated with radioactive tritium and phosphorescent paint. During lunch, brother Bob has some ice cream. The amount of air whipped into that ice cream was measured by a radioisotopic gauge. After you and your family return home this evening, some of you may have soda and others may sit and relax.

Nuclear science is at work here: The soda bottle was carefully filled — a radiation detector prevented spillover. And your family is safe at home because the ionizing smoke detector, using a tiny bit of americium-241, will keep watch over you while you sleep.

Increasingly, our country has become a nation of electricity users. We depend on an abundant affordable supply of energy to power the many machines we use in our complex society. Can you imagine what it would be like not having electricity in your home? About one-third of our energy resources are used to produce electricity.

Electricity can be produced in many ways — most of which you already know about. Today, we're going to talk about one of those ways — nuclear fission. In America, nuclear energy plants are the second largest source of electricity after coal — producing approximately 21 per cent of our electricity. There is something I want all of you to be aware of: The purpose of a nuclear power plant is to produce electricity.

While nuclear power plants have many similarities to other types of electricity generating plants, there are some significant differences.

With the exception of solar, wind, and hydroelectric plants, all others including nuclear convert water to steam that spins the propeller-like blades of a turbine that spins the shaft of a generator. Inside the generator coils of wire and magnetic fields interact to create electricity.

The energy needed to boil water into steam is produced in one of two ways: by burning coal, oil, or gas (fossil fuels) in a furnace or by splitting certain atoms of uranium in a nuclear energy plant. Nothing is burned or exploded in a nuclear energy plant. Rather, the uranium fuel generates heat through a process called fission.

Uranium is an element that can be found in the crust of the Earth. This element, quite abundant in many areas of the world, is naturally radioactive. Certain isotopes of uranium can be used as fuel in a nuclear power plant. The uranium is formed into ceramic pellets about the size of the end of your finger. These pellets are inserted into long, vertical tubes (fuel rods) within the reactor. The reactor is the heart of the nuclear power plant. Basically, it is a machine that heats water.

A reactor has four main parts: the uranium fuel assemblies, the control rods, the coolant/moderator, and the pressure vessel. The fuel assemblies, control rods, and coolant/moderator make up what is known as the reactor core. The core is surrounded by the pressure vessel.

We also have to understand that uranium cannot just be thrown into a reactor the way we shovel coal into a furnace. The fuel rods, containing the uranium, are carefully bound together into fuel assemblies, each of which contains about 240 rods. The assemblies hold the rods apart so that when they are submerged into the reactor core, water can flow between them.

When the uranium atom splits, it releases energy and two or more neutrons from its nucleus. These neutrons can then hit the nuclei of other uranium atoms causing them to fission. The sequence of one fission triggering others, and those triggering still more, is called a chain reaction. When the atoms split, they release energy in the form of heat. The heat is transferred from the reactor core to the steam generator. Here, it's converted to high pressure steam which turns the turbine in the electric generator.

This next transparency shows the basic design of the nuclear fuel steam plant and its various components.

In the past few minutes I've introduced several new terms. Let me explain what each means in the energy generating process. First, control rods. What are they? How are they used?

The control rods slide up and down in between the fuel assemblies in the reactor core. They control or regulate the speed of the nuclear reaction by absorbing neutrons. Here's how it works: When the control rods absorb neutrons, fewer neutrons hit the uranium atoms thus slowing down the chain reaction.

On the other hand, when the core temperature goes down, the control rods are slowly lifted out of the core, and fewer neutrons are absorbed. Therefore, more neutrons are available to cause fission. This releases more heat energy.

Just as there are different types of houses and cars, there are different types of nuclear power plants that generate electricity. The two basic types being used are the boiling water reactor (BWR) and the pressurized water reactor (PWR). These power plants are often referred to as light water reactors.

BOILING WATER REACTOR (BWR)

The boiling water reactor operates in essentially the same way as a fossil fuel generating plant. Neither of these types of power plants have a steam generator. Instead, water in the BWR boils inside the pressure vessel and the steam water mixture is produced when very pure water (reactor coolant) moves upward through the core absorbing heat.

The water boils and produces steam. When the steam rises to the top of the pressure vessel, water droplets are removed, the steam is sent to the turbine generator to turn the turbine.

PRESSURIZED WATER REACTOR (PWR)

The pressurized water reactor differs from the BWR in that the steam to run the turbine is produced in a steam generator. Water boils at 212°F or 100°C. If a lid is tightly placed over a pot of boiling water (a pressure cooker), the pressure inside the pot will increase because the steam cannot escape. As the pressure increases, so does the temperature of the water in the pot. In the PWR plant, a pressurizer unit keeps the water that is flowing through the reactor vessel under very high pressure to prevent it from boiling. The hot water then flows into the steam generator where it is converted to steam. The steam passes through the turbine which produces electricity. About two-thirds of the reactor power plants in the U.S. are of the PWR type.

MEDICAL USES OF RADIOACTIVITY

Since the discovery of radiation, people have benefited from the use of radiation in medicine, agriculture and industry. Physicians use x-rays in more than half of all medical diagnoses to determine the extent of disease or physical injury.

The radioactive isotope cobalt-60 helps to stop the body's immune reaction to transplanted human organs. Also, tests using nuclear materials in hospital laboratories can detect thyroid underactivity in newborn babies. This makes prompt treatment possible, saving many children from mental retardation.

The discovery of x-rays in 1895 was a major turning point in diagnosing diseases because physicians finally had an easy way to "see" inside the body without having to operate. Newer x-ray technologies such as CT (computerized tomography) scans have revolutionized the diagnosis and treatment of diseases affecting almost every part of the body. Other sophisticated techniques have provided physicians with low-risk ways to diagnose heart disease. For example, doctors can now pinpoint cholesterol deposits that are narrowing or blocking coronary arteries, information essential for bypassing or unclogging them.

Every major hospital in the United States has a nuclear medicine department. Radionuclides are used safely and effectively to diagnose and treat a wide variety of diseases more effectively and safely by "seeing" how the disease process alters the normal function of an organ.

To obtain this information a patient swallows, inhales, or receives an injection of a tiny amount of a radionuclide. Special cameras reveal where the radioactivity accumulates briefly in the body, providing, for example, an image of the heart that shows normal and malfunctioning tissue.

Radionuclides are also used in laboratory tests to measure important substances in the body, such as thyroid hormone. Radionuclides are used to effectively treat patients with thyroid diseases, including Graves disease, one of the most common forms of hyperthyroidism, and thyroid cancer.

The use of ionizing radiation has led to major improvements in the diagnosis and treatment of patients with cancer. These innovations have resulted in increased survival rates and improved quality of life. Mammography can detect breast cancer at an early stage when it may be curable.

Needle biopsies are more safe, accurate, and informative when guided by x-ray or other imaging techniques. Radiation is used in monitoring the response of tumors to treatment and in distinguishing malignant tumors from benign ones. Bone and liver scans can detect cancers that have spread to these organs.

Half of all people with cancer are treated with radiation, and the number of those who have been cured continues to rise. There are now tens of thousands of individuals alive and cured from various cancers as a result of radiotherapy. In addition, there are many patients who have entered remission as a result of radiotherapy. Radionuclides are also being used to decrease or eliminate the pain associated with cancer, such as that of the prostate or breast cancer that has spread to the bone.

Radionuclides are a technological backbone for much of the biomedical research being done today. They are used in identifying and learning how genes work. Much of the research on AIDS is dependent upon the use of radionuclides. Scientists are also "arming" monoclonal antibodies, antibodies that are produced in the laboratory and engineered to bind to a specific protein on a patient's tumor cells, with radionuclides.

When such "armed" antibodies are injected into a patient, they bind to the tumor cells, which are then killed by the attached radioactivity, but the nearby normal cells are spared. So far, this approach has produced encouraging success in treating patients with leukemia.

Another clinical and research tool, PET scanning (positron emission tomography), involves injecting radioactive material into a person to "see" the metabolic activity and circulation in a living brain. PET studies have enabled scientists to pinpoint the site of brain tumors or the source of epileptic activity, and to better understand many neurological diseases. For example, researchers were able to learn how dopamine, the chemical messenger (neurotransmitter) that is involved in Parkinson's disease, is used by the brain.

RADIATION IN MEDICINE

X-rays are a type of radiation that can pass through our skin. Our bones are denser than our skin, so when x-rayed, bones and other dense materials cast shadows that can be detected on photographic film.

The effect is similar to placing a pencil behind a piece of paper and holding them in front of a light. The shadow of the pencil is revealed because most light has enough energy to pass through the paper, while the denser pencil stops all the light. The difference is that we need film to see the x-rays for us.

Today, doctors and dentists use x-rays to see structures inside our bodies. This allows them to spot broken bones and dental problems. X-ray machines have now been connected to computers in the development of machines called CAT scanners. These instruments provide doctors with colour TV pictures that show the shape of internal organs.

Approximately 10 million nuclear medicine procedures are performed in the United States annually. Diagnostic x-rays and or radiation therapy were administered to about seven out of every 10 Americans. Medical procedures using radiation have saved thousands of lives through the detection and treatment of conditions ranging from hyperthyroidism to bone cancer.

In such procedures, doctors administer slightly radioactive substances to patients, which are attracted to certain internal organs such as the pancreas, kidney, thyroid, liver, or brain, to diagnose clinical conditions. Moreover, radiation is often used to treat certain types of cancer. Radioactive iodine, specifically iodine-131, is being used frequently to treat thyroid cancer, a disease which strikes about 11,000 Americans every year

USES OF RADIATION: FOOD IRRADIATION

Food irradiation involves exposing the food, either packaged or in bulk, to carefully controlled amounts of ionizing radiation for a specific time to achieve certain desirable objectives. The process cannot increase the normal radioactivity level of the food, regardless of how long the food is exposed to the radiation, or how much of an energy "dose" is absorbed.

It can prevent the division of living cells, such as bacteria and cells of higher organisms, by changing their molecular structure. It can also slow down ripening or maturation of certain fruits and vegetables by causing biochemical reactions in physiological processes of plant tissues.

FOOD PRESERVATION METHODS

Preservation methods for foods have been known and used for centuries: drying, smoking, curing, and using various simple chemical treatments such as salt, are widespread but they all suffer from the problem that they change the way food looks and tastes.

More recently, refrigeration has emerged as the most widely used means of preserving food, or more strictly, for delaying or inhibiting the natural decomposition processes that occur as a result of bacterial action. It also has the advantage that it has a minimal effect on the taste and consistency of food but it is, relatively speaking, an expensive process.

The modern food industry has to make certain choices as to how and when it treats food during the food production cycle. It can start by reducing the level of micro organisms and pests in food by using chemical treatments and pesticides during growth.

For this to be effective the food must then be secured against fresh contamination during transport and storage. An alternative approach is to do very little to the food as grown and harvested, but to treat it nearer to the point of consumption. This is, for example, common with herbs and spices. The food industry will, not surprisingly, tend to choose the way it deals with contamination on the basis of the economics of each case. It will balance the losses in storage and costs of post-treatment against the costs and problems of using more hygienic production and storage methods in the country of origin.

Even where food is produced relatively close to the point of consumption, it may have to be treated because contamination is inherent in the production process. This is why milk has to be pasteurised. Pasteurisation is the most effective way of killing micro organisms with minimal effect on the food itself. Unfortunately pasteurisation can only be used on a very limited range of foods.

NECESSICITY FOOD IRRADIATION

Presently over 40 countries have approved applications to irradiate approximately 40 different foods. These include such items as fruits, vegetables, spices, grains, seafood, meat and poultry. More than half a million tonnes of food is now irradiated throughout the world on a yearly basis. Although this amount represents only a fraction of the food consumed annually, it is constantly growing. This trend is due to three main factors:

- Increasing concerns over food borne diseases.
- High food losses from infestation, contamination and spoilage.
- Growing international trade in food products.

Increasing Concerns Over Foodborne Diseases

Foodborne diseases pose a widespread threat to human health and they are an important cause of reduced economic productivity. Studies by the US Centre for Disease Control in 1999 estimated that foodborne diseases cause approximately 76 million illnesses, 325,000 hospitalizations, and 5,000 deaths in the United States each year.

Economic losses associated with such foodborne diseases are high-estimated between US \$6.5 billion and \$33 billion. The World Health Organization (WHO) has said that food borne diseases are perhaps the most widespread health problem in the contemporary world and an important cause of reduced economic productivity. WHO has also reported that Europe is suffering from an epidemic of salmonellosis.

Pregnant women in the UK have been warned not to eat unpasteurised cheeses which now may have high levels of listeria monocytogenes. Furthermore, even in relatively advanced countries like the UK, the authorities have admitted that the contamination problem is out of control: "The multiplicity of potential routes of contamination makes the elimination of microbiological contamination from poultry being presented for slaughter a virtual impossibility.

This need not be the case, however, as has been demonstrated in Sweden. There it has taken 20 years of ruthless killing of any flock with a salmonella infection to achieve 99% of flocks free of salmonella. Poultry costs more as a result but the Swedish authorities and consumers clearly believe this is worth paying.

High Food Losses from Infestation, Contamination and Poilage

The UN Food and Agricultural Authority (FAO) has estimated that about 25% of all worldwide food production is lost after harvesting to insects, bacteria and spoilage.

Economic losses due to insects and microbes have been estimated to fall between $5 and $17 billion yearly in the US alone. Food irradiation can help reduce these losses and can also reduce our dependence on chemical pesticides, some of which are extremely harmful to the environment (e.g. methyl bromide).

Growing International Trade in Food Products

As our economies become more global, food products must meet high standards of quality and quarantine in order to move across borders. Irradiation is an important tool in the fight to prevent the spread of deleterious insects and microorganisms.

AFFECT OF IRRADIATION ON FOOD

The process involves exposing the food, either packaged or in bulk, to carefully controlled amounts of ionizing radiation for a specific time to achieve certain desirable objectives. One objective is the reduction of microorganisms. When microbes present in the food are irradiated, the energy from the radiation breaks the bonds in the DNA molecules, causing defects in the genetic instructions.

Unless this damage can be repaired, the organism will die or will be unable to reproduce. It matters if the food is frozen or fresh, because it takes larger radiation dose to kill microbes in frozen foods. The effectiveness of the process depends also on the organism's sensitivity to irradiation, on the rate at which it can repair damaged DNA, and especially on the amount of DNA in the target organism:

- Parasites and insect pests, which have large amounts of DNA, are rapidly killed by an extremely low dose of irradiation.
- It takes more irradiation to kill bacteria, because they have less DNA.
- Viruses are the smallest pathogens that have nucleic acid, and they are, in general, resistant to irradiation at doses approved for foods.

Another objective is to delay ripening and sprouting. If the food contains living cells, they will be damaged or killed just as microbes are. This is a useful effect: it can be used to prolong the shelf life of fruits and vegetables because it inhibits sprouting and delays ripening.

IRRADIATION COMPARED TO PASTEURIZATION

As in the heat pasteurization of milk, the irradiation process greatly reduces but does not eliminate all bacteria. Irradiated poultry, for example, still requires refrigeration, but would be safe longer than untreated poultry. Strawberries that have been irradiated will last two to three weeks in the refrigerator compared to only a few days for untreated berries. Irradiation complements, but does not replace, the need for proper food handling practices by producers, processors, and consumers.

A big advantage of irradiated food, is that it is a cold process: the food is still essentially "raw", because it hasn't undergone any thermal process and it does not significantly increase the temperature or change the physical or sensory characteristics of most foods. An irradiated apple, for example, will still be crisp and juicy. Fresh or frozen meat can be irradiated without cooking it.

APPLICATIONS FOR FOOD IRRADIATION

Because the irradiation process works with both large and small quantities, it has a wide range of potential uses. For example, a single serving of poultry can be irradiated for use on a space flight. Or, a large quantity of potatoes can be treated to reduce sprouting during warehouse storage. However, irradiation cannot be used with all foods. It causes undesirable flavour changes in dairy products, for example, and it causes tissue softening in some fruits, such as peaches and nectarines (due to breaking of bonds in cellulose and starch). Irradiation is most useful in four areas:

Preservation

Irradiation can be used to destroy or inactivate organisms that cause spoilage and decomposition, thereby extending the shelf life of foods. It is an energy-efficient food preservation method that has several advantages over traditional canning. The resulting products are closer to the fresh state in texture, flavour, and colour. Using irradiation to preserve foods requires no additional liquid, nor does it cause the loss of natural juices. Both large and small containers can be used and food can be irradiated after being packaged or frozen.

Sterilization

Foods that are sterilized by irradiation can be stored for years without refrigeration—just like canned (heat sterilized) foods. With irradiation it will be possible to develop new shelf-stable products. Sterilized food is useful in hospitals for patients with severely impaired immune systems, such as some patients with cancer or AIDS. These foods can be used by the military and for space flights.

Control Sprouting, Ripening, and Insect Damage

In this role, irradiation offers an alternative to chemicals for use with potatoes, tropical and citrus fruits, grains, spices, and seasonings. However, since no residue is left in the food, irradiation does not protect against reinfestation like insect sprays and fumigants do.

Control Food Borne Illness

Irradiation can be used to effectively eliminate those pathogens that cause foodborne illness, such as Salmonella.

APPROVED FOR IRRADIATION

A variety of foods have been approved for irradiation, for several different purposes. For meats, separate approval is required both from the FDA and the USDA.

Aroval	Food	Purpose
1963	Wheat flour	Control of mold
1964	White Potatoes	Inhibit Sprouting
1986	Pork	Kill Trichina Parasites
1986	Fruit and vegetables	Insect Control
		Increase shelf life
1986	Herbs and spices	Sterilization
1990-FDA	Poultry	Bacterual Pathogen reduction
1992-USDA		
1997-FDA	Meat	Bacterial Pathogen Reduction

Approval	Food	Purpose
1963	Wheat flour	Control of mold
1964	White Potatoes	Inhibit Sprouting
1986	Pork	Kill Trichina Parasites
1986	Fruit and vegetables	Insect Control
		Increase shelf life
1986	Herbs and spices	Sterilization
1990-FDA	Poultry	Bacterial Pathogen Reduction
1992-USDA		
1997-FDA	Meat	Bacterial Pathogen Reduction
1999-USDA		

However, irradiated foods are not widely available yet. Some stores have sold irradiated fruits and vegetables since the early 1990s. Irradiated poultry is available in some grocery stores— mostly small, independent markets— and on menus of a few restaurants. On the other hand, most spices sold wholesale in this country are irradiated, which eliminates the need for chemical fumigation to control pests.

American astronauts have eaten irradiated foods in space since the early 1970s. Patients with weakened immune systems are sometimes fed irradiated foods to reduce the chance of a life-threatening infection. In addition, irradiation is widely used to sterilize a variety of medical and household products, such as joint implants, band-aids, baby pacifiers, cosmetic ingredients, wine and bottle corks, and food packaging materials.

Table. Uses of Radioactive Substances.

Isotope	Use
Carbon-14	Radiometric dating: determination of age of carbon-containing artifacts up to about 70,000 years also used as a biological tracer, for example, in studies of photosynthesis. Naturally occurring radioisotope.
Chlorine-36	Measurement of sources of chloride and determining the age of water up to about 2 million years old. Naturally occurring radioisotope.
Lead-210	Date layers of soil and sand deposited up to about 80 years ago. Naturally occurring radioisotope
Tritium	Measure the age of 'young' groundwater up to about 30 years old. Naturally occurring radioisotope. Titriated water is used to study sewage and liquid wastes.

Potassium/	Radiometric dating:100 000 to several billion years
Argon	Rubidium/Strontium radiometric dating:millions of years
Uranium/Lead	Radiometric dating:millions to billions of years
Oxygen-18	Biological tracer, for example, in studies of photosynthesis
Sodium-24	location of leaks in water pipes, studies of body electrolytes. Isotope prepared in a nuclear reactor
Magnesium-27	Location of leaks in water pipes
Potassium-42	Determination of exchanged potassium in blood flow. Isotope prepared in a nuclear reactor
Chromium-51	Labelling of red blood cells & quantifying gastro-intestinal protein loss. Isotope prepared in a nuclear reactor
Iron-59	In blood studies, when incorporated into steel it is used to determine the amount of friction in machinery
Cobalt-60	Cancer treatment as tumour cells tend to be more susceptible to radiation than other cells
Gallium-67	Tumour-seeking agent. Isotope prepared in a cyclotron
Krypton-81	Lung ventilation studies. Isotope prepared in a cyclotron
Technetium-99	Medical tracer used to locate brain tumours and problems with the lungs, thyroid, liver, spleen, kidney, gall bladder, skeleton, blood pool, bone marrow, salivary & lacrimal glands & heart blood pool & to detect infection. Isotope prepared in a nuclear reactor
Iodine-131	Medical tracer to study & treat the thyroid gland & used in the diagnosis of adrenal medullary & for imaging suspected neural crest and other endocrine tumours. Isotope prepared in a nuclear reactor
Iodine-123	Used in imaging to monitor thyroid function & detect adrenal dysfunction. Isotope prepared in a cyclotron
Ytterbium-169	Brain scans. Isotope prepared in a nuclear reactor
Uranium-235	Enriched as a fuel for most nuclear reactors
Plutonium-239	Fuel for most "fast-breeder" nuclear reactors
Americium-241	Domestic smoke alarms & neutron gauging. Americium-241 is a decay product of plutonium-241 formed in nuclear reactors.
Copper-64	Studying genetic disease affecting copper metabolism. Isotope prepared in a nuclear reactor
Iridium-192	Supplied as a wire for use as an internal radiotherapy device

Molybdenum-99	Used as the 'parent' in a generator to produce technetium-99m, the most widely used radioisotope in nuclear medicine. Isotope prepared in a nuclear reactor
Phosphorus-32	Treatment of excess red blood cells. Isotope prepared in a nuclear reactor
Samarium-153	Used in the treatment of pain associated with bony metastases of primary tumours. Isotope prepared in a nuclear reactor
Yttrium-90	Liver cancer therapy. Isotope prepared in a nuclear reactor
Thallium-201	Locating damaged heart muscle. Isotope prepared in a cyclotron
Caesium-137	Radiotracing to identify sources of soil erosion & depositing, also used in thickness gauging
Gold-198	Trace factory waste causing ocean pollution and to trace sand movement in river beds and on ocean floors
Gadolinium	Used in X-ray fluorescence
Zinc-65	Used to predict the behaviour of heavy metals in
Manganese-54	Effluents from mining waste water

Chapter 16

Radiography

Radiography is the use of X-rays to view unseen or hard-to-image objects. The main diagnostic purposes of X-rays are to see inside ones' body, especially of the brain and fetus, where the bones can be viewed at an optimum resolution (128 shades of grey).

The impact on society of this technique has also been immense. Physicists and researchers have developed numerous types of medicines to assist in the function and development of organs such as the brain and heart, and radiography has improved the economy of most countries in the western world due to the employment of physicists and doctors.

Radiography is used for both medical and industrial applications. If the object being examined is living, whether human or animal, it is regarded as medical; all other radiography is regarded as industrial radiographic work.

HISTORY OF RADIOGRAPHY

Fig. Taking an X-ray Image with Early Crookes Tube Apparatus, Late 1800s.

Radiography started in 1895 with the discovery of X-rays, also referred to as Röntgen rays after Wilhelm Conrad Röntgen who first described their properties in rigorous detail. These previously unknown rays (hence the X) were found to be a type of electromagnetic radiation. It wasn't long before X-rays were used in various applications, from helping to fit shoes, to the medical uses that have persisted.

X-rays were put to diagnostic use very early, before the dangers of ionizing radiation were discovered. Indeed Madam Curie pushed for radiography to be used to treat wounded soldiers in World War I. Initially, many kinds of staff conducted radiography in hospitals, including physicists, photographers, doctors, nurses, and engineers. The medical specialty of radiology grew up over many years around the new technology.

When new diagnostic tests involving X-rays were developed, it was natural for the radiographers to be trained in and to adopt this new technology. This happened first with fluoroscopy, computed tomography , mammography, ultrasound, and magnetic resonance imaging. Although a nonspecialist dictionary might define radiography quite narrowly as "taking X-ray images", this has long been only part of the work of "X-ray departments", radiographers, and radiologists.

EQUIPMENT

A number of sources of X-ray photons have been used; these include sealed X-ray tubes, betatrons, and linear accelerators (linacs). For gamma rays, radioactive sources such as ^{192}Ir have been used.

Detectors

A range of detectors including photographic film, scintillator and semiconductor diode arrays have been used to collect images.

THEORY OF X-RAY ATTENUATION

X-ray photons used for medical purposes are formed by an event involving an electron, while gamma ray photons are formed from an interaction with the nucleus of an atom. In general, medical radiography is done using X-rays formed in an X-ray tube. Nuclear medicine typically involves gamma rays.

The types of electromagnetic radiation of most interest to radiography are X-ray and gamma radiation. This radiation is much more energetic than the more familiar types such as radio waves and visible light. It is this relatively high energy which makes gamma rays useful in radiography but potentially hazardous to living organisms.

The radiation is produced by X-ray tubes, high energy X-ray equipment or natural radioactive elements, such as radium and radon, and artificially produced radioactive isotopes of elements, such as cobalt-60 and iridium-192. Electromagnetic radiation consists of oscillating electric and magnetic fields, but is generally depicted as a single sinusoidal wave. While in the past radium and radon have both been used for radiography, they have fallen out of use as they are radiotoxic alpha radiation emitters which are expensive; iridium-192 and cobalt-60 are far better photon sources.

Such a wave is characterised by its wavelength (the distance from a point on one cycle to the corresponding point on the next cycle) or its frequency (the number of oscillations per second). In a vacuum, all electromagnetic waves travel at the same speed, the speed of light (*c*). The wavelength (ë, lambda) and the frequency (f) are all related by the equation:

$$f = \frac{c}{\lambda}$$

This is true for all electromagnetic radiation.

Electromagnetic radiation is known by various names, depending on its energy. The energy of these waves is related to the frequency and the wavelength by the relationship:

$$E = hf = h\left(\frac{c}{\lambda}\right)$$

where *h* is a constant known as the Planck constant.

Gamma rays are indirectly ionizing radiation. A gamma ray passes through matter until it undergoes an interaction with an atomic particle, usually an electron. During this interaction, energy is transferred from the gamma ray to the electron, which is a directly ionizing particle. As a result of this energy transfer, the electron is liberated from the atom and proceeds to ionize matter by colliding with other electrons along its path.

Other times, the passing gamma ray interferes with the orbit of the electron, and slows it, releasing energy but not becoming dislodged. The atom is not ionised, and the gamma ray continues on, although at a lower energy.

This energy released is usually heat or another, weaker photon, and causes biological harm as a radiation burn. The chain reaction caused by the initial dose of radiation can continue after exposure, much like a sunburn continues to damage skin even after one is out of direct sunlight.

For the range of energies commonly used in radiography, the interaction between gamma rays and electrons occurs in two ways. One effect takes place where all the gamma ray's energy is transmitted to an entire atom. The gamma ray no longer exists and an electron emerges from the atom with kinetic (motion in relation to force) energy almost equal to the gamma energy. This effect is predominant at low gamma energies and is known as the photoelectric effect. The other major effect occurs when a gamma ray interacts with an atomic electron, freeing it from the atom and imparting to it only a fraction of the gamma ray's kinetic energy. A secondary gamma ray with less energy (hence lower frequency) also emerges from the interaction. This effect predominates at higher gamma energies and is known as the Compton effect.

In both of these effects the emergent electrons lose their kinetic energy by ionizing surrounding atoms. The density of ions so generated is a measure of the energy delivered to the material by the gamma rays.

The most common means of measuring the variations in a beam of radiation is by observing its effect on a photographic film. This effect is the same as that of light, and the more intense the radiation is, the more it darkens, or exposes, the film.

Other methods are in use, such as the ionizing effect measured electronically, its ability to discharge an electrostatically charged plate or to cause certain chemicals to fluoresce as in fluoroscopy.

MEDICAL RADIOGRAPHY

Radiography is the use of ionizing electromagnetic radiation such as X-rays to view objects. Although not technically radiographic techniques, imaging modalities such as PET and MRI are sometimes grouped in radiography due to the fact that the radiology department of hospitals handle all forms of imaging. Treatment using radiation is known as radiotherapy.

Radiography started in 1895 with the discovery of X-rays (later also called Röntgen rays after the man who first described their properties in rigorous detail), a type of electromagnetic radiation. Soon these found various applications, from helping to find shoes that fit, to the more lasting medical uses. X-rays were put to diagnostic use very early, before the dangers of ionising radiation were discovered. Initially, many groups of staff conducted radiography in hospitals, including physicists, photographers, doctors, nurses, and engineers.

The medical speciality of radiology grew up around the new technology, and this lasted many years. When new diagnostic tests involving X-rays were developed, it was natural for the radiographers to be trained and adopt this new technology. This happened first with fluoroscopy, computed tomography, and mammography.

Ultrasound and magnetic resonance imaging was added to the list of skills used by radiographers because they are also medical imaging, but these disciplines do not use ionising radiation or X-rays. Although a nonspecialist dictionary might define radiography quite narrowly as "taking X-ray images", this has only been part of the work of an "X-ray department", radiographers, and radiologists for a very long time. X-rays are also exploited by industrial radiographers in the field of nondestructive testing, where the newer technology of ultrasound is also used.

DIAGNOSTIC RADIOGRAPHY

Diagnostic radiography involves the use of both ionising radiation and non-ionising radiation to create images for medical diagnoses. The predominant test is still the X-ray (the word X-ray is often used for both the test and the actual film or digital image). X-rays are the second most commonly used medical tests, after laboratory tests. This application is known as diagnostic radiography. Since the body is made up of various substances with differing densities, X-rays can be used to reveal the internal structure of the body on film by highlighting these differences using attenuation, or the *absorption* of X-ray photons by the denser substances (like calcium-rich bones). Medical diagnostic radiography is undertaken by a specially trained professional called a diagnostic radiographer in the UK, or a radiologic technologist in the USA.

Projection Radiography

The creation of images by exposing an object to X-rays or other high-energy forms of electromagnetic radiation and capturing the resulting remnant beam (or "shadow") as a latent image is known as "projection radiography." The "shadow" may be converted to light using a fluorescent screen, which is then captured on photographic film, it may be captured by a phosphor screen to be "read" later by a laser (CR), or it may directly activate a matrix of solid-state detectors (DR—similar to a very large version of a CCD in a digital camera).

Bone and some organs (such as lungs) especially lend themselves to projection radiography. It is a relatively low-cost investigation with a high diagnostic yield.

Projection radiography uses X-rays in different amounts and strengths depending on what body part is being imaged:

- Hard tissues such as bone require a relatively high energy photon source, and typically a tungsten anode is used with a high voltage (50-150 kVp) on a 3-phase or high-frequency machine to generate braking radiation. Bony tissue and metals are denser than the surrounding tissue, and thus by absorbing more of the X-ray photons they prevent the film from getting exposed as much. Wherever dense tissue absorbs or stops the X-rays, the resulting X-ray film is unexposed, and appears translucent blue, whereas the black parts of the film represent lower-density tissues such as fat, skin, and internal organs, which could not stop the X-rays. This is usually used to see bony fractures, foreign objects (such as ingested coins), and used for finding bony pathology such as osteoarthritis, infection (osteomyelitis), cancer (osteosarcoma), as well as growth studies (leg length, achondroplasia, scoliosis, etc).

- Soft tissues are seen with the same machine as for hard tissues, but a "softer" or less-penetrating X-ray beam is used. Tissues commonly imaged include the lungs and heart shadow in a chest X-ray, the air pattern of the bowel in abdominal X-rays, the soft tissues of the neck, the orbits by a skull X-ray before an MRI to check for radiopaque foreign bodies (especially metal), and of course the soft tissue shadows in X-rays of bony injuries are looked at by the radiologist for signs of hidden trauma (for example, the famous "fat pad" sign on a fractured elbow).
- Dental radiography uses a small radiation dose with high penetration to view teeth, which are relatively dense. A dentist may examine a painful tooth and gum using X-ray equipment. The machines used are typically single-phase pulsating DC, the oldest and simplest sort. Dental technicians or the dentist may run these machines— radio logic technologists are not required by law to be present.
- Mammography is an X-ray examination of breasts and other soft tissues. This has been used mostly on women to screen for breast cancer, but is also used to view male breasts, and used in conjunction with a radiologist or a surgeon to localise suspicious tissues before a biopsy or a lumpectomy. Breast implants designed to enlarge the breasts reduce the viewing ability of mammography, and require more time for imaging as more views need to be taken. This is because the material used in the implant is very dense compared to breast tissue, and looks white (clear) on the film. The radiation used for mammography tends to be softer (has a lower photon energy) than that used for the harder tissues. Often a tube with a molybdenum anode is used with about 30 000 volts (30 kV), giving a range of X-ray energies of about 15-30 keV. Many of these photons are "characteristic radiation" of a specific energy determined by the atomic structure of the target material (Mo-K radiation).

Other modalities are used in radiography when traditional projection X-ray cannot image what doctors want to see. Below are other modalities included within radiography; they are only summaries and more specific information can be viewed by going to their individual pages:

Fluoroscopy (Angiography, Gastro-intestinal Fluoroscopy)

Fluoroscopy is a term invented by Thomas Edison during his early

X-ray studies. The name refers to the fluorescence he saw while looking at a glowing plate bombarded with X-rays.

This is a technique that provides moving projection radiographs of lower quality. Fluoroscopy is mainly performed to view movement (of tissue or a contrast agent), or to guide a medical intervention, such as angioplasty, pacemaker insertion, or joint repair/replacement. The latter are often carried out in the operating theatre, using a portable fluoroscopy machine called a C-arm. It can move around the surgery table and make digital images for the surgeon.

Angiography is the use of fluoroscopy to view the cardiovascular system. An iodine-based contrast is injected into the bloodstream and watched as it travels around. Since liquid blood and the vessels are not very dense, a contrast with high density (like the large iodine atoms) is used to view the vessels under X-ray. Angiography is used to find aneurysms, leaks, blockages (thromboses), new vessel growth, and placement of catheters and stents. Balloon angioplasty is often done with angiography.

Fluoroscopy can be used to examine the digestive system using a substance which is opaque to X-rays, (usually barium sulfate or gastrografin), which is introduced into the digestive system either by swallowing or as an enema. This is normally as part of a *double contrast technique,* using positive and negative contrast.

Barium sulfate coats the walls of the digestive tract (positive contrast), which allows the shape of the digestive tract to be outlined as white or clear on an X-ray. Air may then be introduced (negative contrast), which looks black on the film.

The barium meal is an example of a contrast agent swallowed to examine the upper digestive tract. Note that while soluble barium compounds are very toxic, the insoluble barium sulfate is non-toxic because its low solubility prevents the body from absorbing it.

- *A number of substances have been used as positive contrast agents*: silver, bismuth, caesium, thorium, tin, zirconium, tantalum, tungsten and lanthanide compounds have been used as contrast agents. The use of thoria (thorium dioxide) as an agent was rapidly stopped as thorium causes liver cancer.

Most modern injected radiographic positive contrast media are iodine-based. Patients who suffer from allergy to shellfish may be allergic to iodine, and should consult their physician regarding pre-medication to lessen risk of allergic reaction. Iodinated contrast comes in two forms: ionic and non-ionic compounds.

Non-ionic contrast is significantly more expensive than ionic (approximately three to five times the cost), however, non-ionic contrast tends to be safer for the patient, causing fewer allergic reactions and uncomfortable side effects such as hot sensations or flushing. Most imaging centers now use non-ionic contrast exclusively, finding that the benefits to patients outweigh the expense.

- Negative radiographic contrast agents are air and carbon dioxide (CO_2). The latter is easily absorbed by the body and causes less spasm. It can also be injected into the blood, where air absolutely cannot.

Dual Energy X-ray Absorptiometry

DEXA, or bone densitometry, is used primarily for osteoporosis tests. It is not projection radiography, as the X-rays are emitted in 2 narrow beams that are scanned across the patient, 90 degrees from each other. Usually the hip (head of the femur), lower back (lumbar spine) or heel (calcaneum) are imaged, and the bone density (amount of calcium) is determined and given a number (a T-score).

It is not used for bone imaging, as the image quality is not good enough to make an accurate diagnostic image for fractures, inflammation etc. It can also be used to measure total body fat, though this isn't common. The radiation dose received from DEXA scans is very low, much lower than projection radiography examinations.

Computed Tomography

Computed tomography or CT scan (previously known as CAT scan, the "A" standing for "axial") uses a high amount of ionizing radiation (in the form of X-rays) in conjunction with a computer to create images of both soft and hard tissues. These images look as though the patient was sliced like bread (thus, "tomography"— "tomo" means "slice").

The machine looks similar to an MRI machine to many patients, but is not related. The exams are generally short, most lasting only as long as a breath-hold. Contrast agents are often used, depending on the tissues needing to be seen.

Radiographers perform these examinations, sometimes in conjunction with a radiologist (for instance, when a radiologist performs a CT-guided biopsy).

TECHNICAL CONSIDERATIONS

X-ray photons are formed in events involving electrons and are the mainly form of ionizing electromagnetic radiation used in medical radiography. This radiation is much more energetic than the more familiar types such as radio waves and visible light. Proper production and detection of photons are important in the creation of good radiograms.

Photon Production

X-ray radiation for medical imaging is typically produced by X-ray tubes, which operate through bombarding the anode with high energy electrons emitted from a hot cathode. Image sharpness, contrast, and patient dosage are important considerations in medical radiography and these requirements determined the desired energies of the tube, the type of material used on the anode, and the method in which the power is generated to drive the tube.

Although the technical definition of x-rays rage from 1-700 keV, medical x-rays use typically of 5-150 keV x-rays. The photons emitted come in discrete bands of energy corresponding to the material of the anode, and the undesired band are removed.

Choice of the anode and its emitted radiation energies depends on the application and the tissues being imaged, for instance molybdenum is often used in mammography due to it 20 keV x-rays. Too high radiation energies will result in poor pictures since the radiation cannot be readily attenuated, however too low energies will increase the radiation dosage of the patient without improvements in image quality.

Sharpness of a radiographic image is strongly determined by the the size of the x-ray source. This is determined by the are of the electron beam hitting the anode. A large photon sources results in more blurring in the final image and is worsened a increase in image formation distance. This blurring can be measured as a contribution to the modulation transfer function of the imaging system.

Power Generation

The power used by the x-ray tube is generated by a specialized generator, which supplies the voltage and current required to drive the tube. The generator needs to supply high voltages with small exposure times. An exposure thus can be described by two factor:

- The *peak voltage* of the cathode to anode
- The *milliamprere seconds* exposure time

These variables can be controlled by the operator but is more typically assigned automatically by the x-ray machinery through sampling the emitted radiation. Power generators convert standard 120 or 220 volt AC to higher DC voltages and typically employ rectified and filtered multiphase transformers which maintain a constant voltage and can be turn rapidly on and off for millisecond exposures.

Photon Detection

Photons images that have been shadowed from an imaging subject must be detected at high fidelity and resolution to allow for diagnosis.

There are three main types of image detection methods used namely: film/ screens, image intensifiers, and digital detectors, with the latter fast becoming the standard for x-ray image detection.

Film/Screens

X-ray film is almost always used in conjunction with x-ray sensitive screen due to the fact that high resolution film is quite poor at detecting x-rays. These screens contain rare earth minerals and phosphor materials that convert x-ray radiation to visible light or lows EM energies to which the film is sensitive.

Screen generally have to have good contrast, dynamic range, and resolution, with the former two factors being competing properties. The resolution of the screen also affects the sensitivity of the detectors since more sensitive screens are generally thicken, which causes the more blurring due to spreading light.

The film speed also play a factor in image quality. Higher speeds are more sensitive to photons but are generally lower in resolution and more susceptible to noise. Lower speed films produce images of good resolution and dynamic range but requires more photons for exposure and increases the radiation dosage of the subject.

Image Intensifiers and Array Detectors

Image intensifters are analog devices that readily convert the acquired x-ray image into one visible on an video screen. This device is made of a vacuum tube with a wide input surface coated on the inside with caesium iodide (CsI).

When hit by x-rays material phosphors which causes the photocathode adjacent to it to emit electrons. These electron are then focus using electron lenses inside the intensifier to a output screen coated with phosphorescent materials. The image from the output can then be recorded via a camera and displayed.

Digital devices known an array detectors are becoming more common in fluoroscopy. These devices are made of discrete pixelated detectors known as TFTs which can either work *indirectly* by using photo detectors that detect light emitted from a scintillator material such as CsI, or *directly* by capturing the electrons produced when the x-rays hit the detector.

Direct detector do not tend to experience the blurring or spreading effect caused by phosphorescent scintillators of or film screens since the detectors are activated directly by x-ray photons.

INDUSTRIAL RADIOGRAPHY

Industrial Radiography is the use of ionising electromagnetic radiation to view objects in a way that can't be seen otherwise. It is not to be confused with the use of ionizing radiation to change or modify objects; radiography's purpose is strictly for viewing.

Industrial radiography has grown out of engineering, and is a major element of nondestructive testing (NDT). It is a method of inspecting materials for hidden flaws by using the ability of short wavelength electromagnetic radiation to penetrate various materials.

Radiography started in 1895 with the discovery of X-rays (later also called Röntgen rays after the man who first described their properties in detail), a type of electromagnetic radiation. Soon after the discovery of X-rays, radioactivity was discovered.

By using radioactive sources such as radium, far higher photon energies could be obtained than those which can be obtained from *normal* X-ray machines. Soon these found various applications, with one of the earliest users being Loughborough University, from helping to fit shoes, more lasting medical uses and the examination of non-living objects.

X-rays and gamma-rays were put to use very early, before the dangers of ionising radiation were discovered. After World War II new isotopes such as caesium-137, iridium-192 and cobalt-60 became available for industrial radiography, and the use of radium and radon decreased.

USES

Inspection of Materials

Gamma radiation sources, most commonly Iridium-192 and Cobalt-60, are used to inspect a variety of materials. The vast majority of radiography concerns the testing and grading of welds on pressurized piping, pressure vessels, high-capacity storage containers, pipelines, and some structural welds.

Other tested materials include concrete (locating rebar or conduit), welder's test coupons, machined parts, plate metal, or pipewall (locating anomalies due to corrosion or mechanical damage). Theoretically, industrial radiographers could radiograph any solid, flat material (walls, ceilings, floors, square or rectangular containers) or any hollow cylindrical or spherical object. For purposes of inspection, including weld inspection, there exist several exposure arrangements.

First, there is the panoramic, one of the four single wall exposure/ single wall view (SWE/SWV) arrangements. This exposure is created when the radiographer places the source of radiation at the centre of a sphere, cone, or cylinder (including tanks, vessels, and piping). Depending upon client requirements, the radiographer would then place film cassettes on the outside of the surface to be examined.

This exposure arrangement is ideal - when properly arranged and exposed, all portions of all exposed film will be of the same approximate density. It also has the advantage of taking less time than other arrangements since the source must only penetrate the total wall thickness (WT) once and must only travel the radius of the inspection item, not its full diameter. The major disadvantage of the panoramic is that it may be impractical to reach the centre of the item (enclosed pipe) or the source may be too weak to perform in this arrangement (large vessels or tanks).

The second SWE/SWV arrangement is an interior placement of the source in an enclosed inspection item without having the source centered up. The source does not come in direct contact with the item, but is placed a distance away, depending on client requirements. The third is an exterior placement with similar characteristics.

The fifth is reserved for flat objects, such as plate metal, and is also radiographed without the source coming in direct contact with the item. In each case, the radiographic film is located on the opposite side of the inspection item from the source. In all four cases, only one wall is exposed, and only one wall is viewed on the radiograph.

Of the other exposure arrangements, only the contact shot has the source located on the inspection item. This type of radiograph will expose both walls, but will only resolve the image of that wall which is nearest the film. This exposure arrangement takes more time than a panoramic, as the source must now penetrate the WT twice and travel the entire outside diameter of the pipe or vessel to reach the film on the opposite side.

This is a double wall exposure/single wall view DWE/SWV arrangement. Another is the superimposure (wherein the source is placed on one side of the item, not in direct contact with it, with the film on the opposite side). This arrangement is usually reserved for very small diameter piping or parts. The last DWE/SWV exposure arrangement is the elliptical, in which the source is offset from the plane of the inspection item (usually a weld in pipe) and the elliptical image of the weld furthest from the source is cast onto the film.

The beam of radiation must be directed to the middle of the section under examination and must be normal to the material surface at that point, except in special techniques where known defects are best revealed by a different alignment of the beam.

The length of weld under examination for each exposure shall be such that the thickness of the material at the diagnostic extremities, measured in the direction of the incident beam, does not exceed the actual thickness at that point by more than 6%.

The specimen to be inspected is placed between the source of radiation and the detecting device, usually the film in a light tight holder or cassette, and the radiation is allowed to penetrate the part for the required length of time to be adequately recorded.

The result is a two-dimensional projection of the part onto the film, producing a latent image of varying densities according to the amount of radiation reaching each area. It is known as a radiograph, as distinct from a photograph produced by light.

Because film is cumulative in its response (the exposure increasing as it absorbs more radiation), relatively weak radiation can be detected by prolonging the exposure until the film can record an image that will be visible after development. The radiograph is examined as a negative, without printing as a positive as in photography. This is because, in printing, some of the detail is always lost and no useful purpose is served.

Before commencing a radiographic examination, it is always advisable to examine the component with one's own eyes, to eliminate any possible external defects. If the surface of a weld is too irregular, it may be desirable to grind it to obtain a smooth finish, but this is likely to be limited to those cases in which the surface irregularities (which will be visible on the radiograph) may make detecting internal defects difficult.

After this visual examination, the operator will have a clear idea of the possibilities of access to the two faces of the weld, which is important both for the setting up of the equipment and for the choice of the most appropriate technique.

Airport Security

Both hold luggage and carry-on hand luggage are normally examined by X-ray machines using X-ray radiography.

Non-intrusive Cargo Scanning

Gamma Radiography and High-Energy X-ray radiography are currently used to scan intermodal freight cargo containers in US and other countries. Also research is being done on adapting other types of radiography like Dual-Energy X-ray Radiography or Muon Radiography for scanning intermodal cargo containers.

SOURCES

X-ray Sources

A high energy X-ray machine can be used. It is often important to use a high accelerating voltage to provide the electrons with a very high energy.

This is because in a braking radiation source the maximum photon energy is determined by the energy of the charged particles. A recent development is the betatron, which is a device similar to a cyclotron that acts as a very intense photon source.

Radioisotope Sources

These have the advantage that they do not need a supply of electrical power to function, but they do have the disadvantage that they can not be turned off. Also it is difficult using radioactivity to create a small and compact source which offers the photon flux which is possible with a normal sealed X-ray tube. It might be possible to use Cs-137 as a photon source for radiography but this isotope is always diluted with inactive caesium isotopes. This makes it difficult to get a physcially small source, and a large volume of the source makes it impossible to capture fine details in a radiographic examination.

Both cobalt-60 and caesium-137 have only a few gamma energies which make them close to monochromatic. The photon energy of cobalt-60 is higher than that of caesium-137, which allows cobalt sources to be used to examine thicker sections of metals than those which could be examined with Cs-137. Iridium-192 has a lower photon energy than cobalt-60 and its gamma spectrum is complex (many lines of very different enegies), but this can be an advantage as this can give better contrast for the final photographs.

It has been known for many years that an inactive iridium or cobalt metal object can be machined to size. In the case of cobalt it is common to alloy it with nickel to improve the mechanical properties. In the case of iridium a thin wire or rod could be used.

These precursor materials can then be placed within stainless steel containers which are leak tested before being converted into radioactive sources. These objects can be processed by neutron activation to form gamma emitting radioisotopes. The stainless steel has only a small ability to be activated and the small activity due to ^{55}Fe and ^{63}Ni are unlikely to pose a problem in the final application because these isotopes are beta emitters which have very weak gamma emission. The ^{59}Fe which might form has a short half life, so by allowing a cobalt source to stand for a year much of this isotope will decay away.

The source is often a very small object which needs to be transported to the site where the work is to be conducted in a shielded container. It is normal to place the film in industrial radiography, clear the area where the work is to be done, add shielding (collimators) to reduce the size of the controlled area before exposing the radioactive source.

A series of different designs have been developed for radiographic "cameras". Rather than the "camera" being a device which accepts photons to record a picture, the "camera" in industrial radiography is the radioactive photon source.

Torch Design of RadioGraphic Cameras

One design is best thought of as being like a torch. The radioactive source is placed inside a shielded box, a hinge allows part of the shielding to be opened exposing the source, allowing photons to exit the radiography camera.

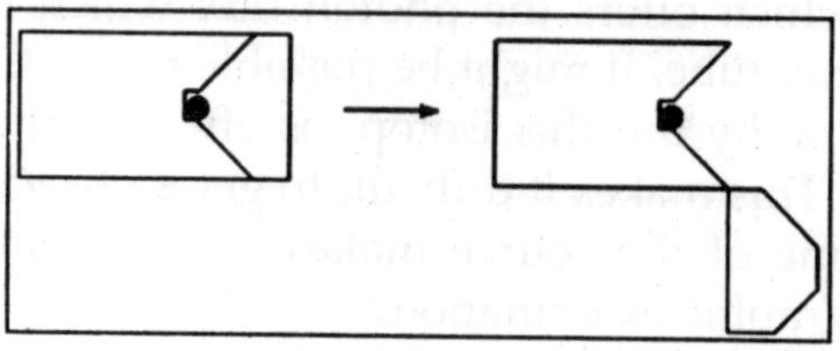

Fig. Torch Type Camera Uses a Hinge.

A torch type camera which uses a hinge. The radioactive source is in red, the shielding by blue/green and the gamma rays are represented by yellow. Another design for a torch is where the source is placed in a metal wheel, which can turn inside the camera to move between the expose and storage positions.

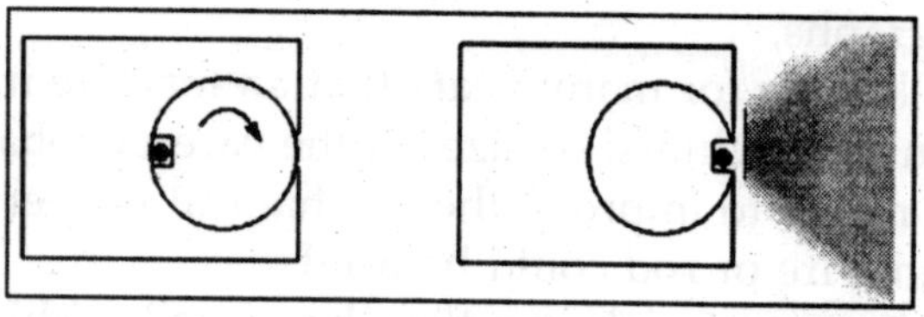

Fig. Torch Type Camera which Uses a Wheel Design.

A torch type camera which uses a wheel design. The radioactive source is in red and the gamma rays are represented by yellow

Cable Based Design of Radiographic Cameras

One group of designs use a radioactive source which comes out on a cable from a shielded container. One such unit was involved in an accident which occurred in Bolivia. This type of radiography could be compared to the remote afterloading method in Brachytherapy.

In one design of equipment the source is stored in a block of lead or DU metal which has a S shaped tube like hole which passes through the block. In the safe position the source is in the centre of the block and is attached to a metal wire which extends in both directions, to use the source a guide tube is attached to one side of the block while a drive cable is attached to the other end of the short cable. Using a hand operated winch the source is then pushed out of the shield and along the guide tube to were it is needed. It was one of these cable-based systems which was involved in an accident which occurred in Bolivia.

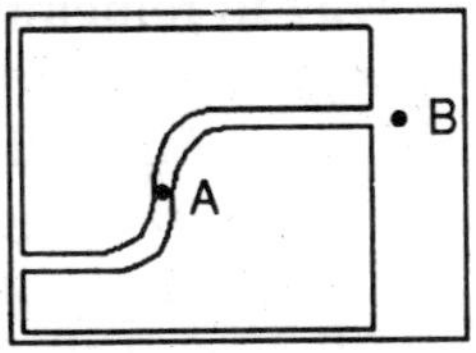

Fig. The S Shaped Hole Passes Through a Metal Block.

A diagram of the S shaped hole which passes through a metal block, the source is stored at point A and it is driven out on a cable through a hole to point B. It will often go a long way along a guide tube to where it is needed.

Contrast Agents

Defects such as delaminations and planar cracks are difficult to detect using radiography, which is why penetrants are often used to enhance the contrast in the detection of such defects. Penetrants used include silver nitrate, zinc iodide, chloroform and diiodomethane. Choice of the penetrant is determined by the ease with which it can penetrate the cracks and also with which it can be removed. Diiodomethane has the advantages of high opacity, ease of penetration, and ease of removal because it evaporates relatively quickly. However, it can cause skin burns.

Neutrons

In some rare cases, radiography is done with neutrons. This type of radiography is called *neutron radiography* (NR, Nray, N-Ray) or *neutron imaging*. Neutron radiography can see very different things than X-rays, because neutrons can pass with ease through lead and steel but are stopped by plastics, water and oils. Neutron sources include radioactive (^{241}Am/Be and Cf) sources, electrically driven D-T reactions in vacuum tubes and conventional critical nuclear reactors. It might be possible to use a neutron amplifier to increase the neutron flux.

Since the amount of radiation emerging from the opposite side of the material can be detected and measured, variations in this amount (or intensity) of radiation are used to determine thickness or composition of material. Penetrating radiations are those restricted to that part of the electromagnetic spectrum of wavelength less than about 10 nanometres.

Safety

Industrial radiographers are in many locations required by governing authorities to use certain types of safety equipment and to work in pairs.

The safety equipment usually includes four basic items: a radiation survey meter (such as a Geiger/Mueller counter), an alarming dosimeter or rate meter, a gas-charged dosimeter, and a film badge or thermoluminescent dosimeter (TLD). The easiest way to remember what each of these items does is to compare them to gauges on an automobile.

The survey meter could be compared to the speedometer, as it measures the speed, or rate, at which radiation is being picked up. When properly calibrated, used, and maintained, it allows the radiographer to see the current exposure to radiation at the meter. It can usually be set for different intensities, and is used to prevent the radiographer from being overexposed to the radioactive source, as well as for verifying the boundary that radiographers are required to maintain around the exposed source during radiographic operations.

The alarming dosimeter could be most closely compared with the tachometer, as it alarms when the radiographer "redlines" or is exposed to too much radiation. When properly calibrated, activated, and worn on the radiographer's person, it will emit an alarm when the meter measures a radiation level in excess of a preset threshold.

This device is intended to prevent the radiographer from inadvertently walking up on an exposed source. The gas-charged dosimeter is like a trip meter in that it measures the total radiation received, but can be reset. It is designed to help the radiographer measure his/her total periodic dose of radiation.

When properly calibrated, recharged, and worn on the radiographer's person, it can tell the radiographer at a glance how much radiation to which the device has been exposed since it was last recharged. Radiographers in many states are required to log their radiation exposures and generate an exposure report. The film badge or TLD is more like a car's odometer. It is actually a specialized piece of radiographic film in a rugged container. It is meant to measure the radiographer's total exposure over time (usually a month) and is used by regulating authorities to monitor the total exposure of certified radiographers in a certain jurisdiction. At the end of the month, the film badge is turned in and is processed. A report of the radiographer's total dose is generated and is kept on file.

When these safety devices are properly calibrated, maintained, and used, it is virtually impossible for a radiographer to be injured by a radioactive overexposure. Sadly, the elimination of just one of these devices can jeopardize the safety of the radiographer and all those who are nearby. Without the survey meter, the radiation received may be just below the threshold of the rate alarm, and it may be several hours before the radiographer checks the dosimeter, and up to a month or more before the film badge is developed to detect a low intensity overexposure.

Without the rate alarm, one radiographer may inadvertently walk up on the source exposed by the other radiographer. Without the dosimeter, the radiographer may be unaware of an overexposure, or even a radiation burn, which may take weeks to result in noticeable injury. And without the film badge, the radiographer is deprived of an important tool designed to protect him or her from the effects of a long-term overexposure to occupationally-obtained radiation, and thus may suffer long-term health problems as a result.

Industrial radiography appears to have one of the worst safety profiles of the radiation professions, possibly because there are many operators using strong gamma sources (> 2 Ci) in remote sites with little supervision when compared with workers within the nuclear industry or within hospitals. Many of the "lost source" accidents commented on by the International Atomic Energy Agency involve radiography equipment. Lost source accidents have the potential to cause a considerable loss of human life.

One scenario is that a passerby finds the radiography source and not knowing what it is, takes it home. The person shortly afterwards becomes ill and dies as a result of the radiation dose. The source remains in their home where it continues to irradiate other members of the household. Such an event occurred in March 1984 in Casablanca (Mohammedia) which is part of Morocco; this is related to the more famous Goiβnia accident, where a related chain of events caused members of the public to be exposed to radiation sources.

X-RAY

X-radiation (composed of X-rays) is a form of electromagnetic radiation. X-rays have a wavelength in the range of 10 to 0.001 nanometers, corresponding to frequencies in the range 30 petahertz to 30 exahertz (30 $\times 10^{15}$ Hz to 30 $\times 10^{18}$ Hz) and energies in the range 120 eV to 120 keV. They are shorter in wavelength than UV rays. In many languages, X-radiation is called Röntgen radiation after one of its first investigators, Wilhelm Conrad Röntgen.

X-rays are primarily used for diagnostic radiography and crystallography. As a result, the term *X-ray* is metonymically used to refer to a radiographic image produced using this method, in addition to the method itself. X-rays are a form of ionizing radiation and as such can be dangerous.

X-rays span 3 decades in wavelength, frequency and energy. From about 0.12 to 12 keV they are classified as soft X-rays, and from about 12 to 120 keV as hard X-rays, due to their penetrating abilities.

The distinction between X-rays and gamma rays has changed in recent decades. Originally, the electromagnetic radiation emitted by X-ray tubes had a longer wavelength than the radiation emitted by radioactive nuclei (gamma rays).

So older literature distinguished between X- and gamma radiation on the basis of wavelength, with radiation shorter than some arbitrary wavelength, such as 10^{-11} m, defined as gamma rays. However, as shorter wavelength continuous spectrum "X-ray" sources such as linear accelerators and longer wavelength "gamma ray" emitters were discovered, the wavelength bands largely overlapped. The two types of radiation are now usually defined by their origin: X-rays are emitted by electrons outside the nucleus, while gamma rays are emitted by the nucleus.

Units of Measure and Exposure

The measure of X-rays ionizing ability is called the exposure:

- The coulomb per kilogram (C/kg) is the SI unit of ionizing radiation exposure, and measures the amount of radiation required to create 1 coulomb of charge of each polarity in 1 kilogram of matter.
- The röntgen (R) is an obsolete older traditional unit of exposure, which represented the amount of radiation required to create 1 esu of charge of each polarity in 1 cubic centimeter of dry air. 1 roentgen = 2.58×10^{-4} C/kg

However, the effect of ionizing radiation on matter (especially living tissue) is more closely related to the amount of energy deposited rather than the charge. This is called the absorbed dose:

- The gray (Gy) which has units of (J/C), is the SI unit of absorbed dose which is the amount of radiation required to deposit 1 joule of energy in 1 kilogram of any kind of matter.
- The rad is the (obsolete) corresponding traditional unit, equal to 0.01 J deposited per kg. 100 rad = 1 Gy.

The equivalent dose is the measure of the biological effect of radiation on human tissue. For X-rays it is equal to the absorbed dose.

- The sievert (Sv) is the SI unit of equivalent dose, which for X-rays is equal to the gray (Gy).
- The rem is the traditional unit of equivalent dose. For X-rays it is equal to the rad or 0.01 J of energy deposited per kg. 1 sievert = 100 rem.

Medical X-rays are a major source of *manmade* radiation exposure, accounting for 58% in the USA in 1987, but since most radiation exposure is natural (82%) it only accounts for 10% of *total* USA radiation exposure.

Reported dosage due to dental X-rays seems to vary significantly. Depending on the source, a typical dental X-ray of a human results in an exposure of perhaps, 3, 40, 300, or as many as 900 mrems (30 to 9,000 ìSv).

Medical Physics

Table. X-ray K-series Spectral Line Wavelengths (nm) for some Common Target Materials.

Target	$K\beta_1$	$K\beta_2$	$K\alpha_1$	$K\alpha_2$
Fe	0.17566	0.17442	0.193604	0.193998
Co				
Ni	0.15001	0.14886	0.165791	0.166175
Cu	0.139222	0.138109	0.154056	0.154439
Zr	0.070173	0.068993	0.078593	0.079015
Mo	0.063229	0.062099	0.070930	0.071359
W				
Re				

X-rays are generated by an X-ray tube, a vacuum tube that uses a high voltage to accelerate electrons released by a hot cathode to a high velocity. The high velocity electrons collide with a metal target, the anode, creating the X-rays. In medical X-ray tubes the target is usually tungsten or a more crack-resistant alloy of rhenium (5%) and tungsten (95%), but sometimes molybdenum for more specialized applications, such as when soft X-rays are needed as in mammography.

In crystallography, a copper target is most common, with cobalt often being used when fluorescence from iron content in the sample might otherwise present a problem.

The maximum energy of the produced X-ray photon in keV is limited by the energy of the incident electron, which is equal to the voltage on the tube, so an 80 kV tube can't create higher than 80 keV X-rays. When the electrons hit the target, X-rays are created by two different atomic processes:

- *X-ray fluorescence*: If the electron has enough energy it can knock an orbital electron out of the inner shell of a metal atom, and as a result electrons from higher energy levels then fill up the vacancy and X-ray photons are emitted. This process produces a discrete spectrum of X-ray frequencies, called spectral lines. The spectral lines generated depends on the target (anode) element used and thus are called characteristic lines. Usually these are transitions from upper shells into K shell (called K lines), into L shell (called L lines) and so on.

- *Bremsstrahlung*: This is radiation given off by the electrons as they are scattered by the strong electric field near the high-Z (proton number) nuclei. These X-rays have a continuous spectrum. The intensity of the X-rays increases linearly with decreasing frequency, from zero at the energy of the incident electrons, the voltage on the X-ray tube.

So the resulting output of a tube consists of a continuous bremsstrahlung spectrum falling off to zero at the tube voltage, plus several spikes at the characteristic lines. The voltages used in diagnostic X-ray tubes, and thus the highest energies of the X-rays, range from roughly 20 to 150 kV.

In medical diagnostic applications, the low energy (soft) X-rays are unwanted, since they are totally absorbed by the body, increasing the dose. So a thin metal (often aluminum) sheet is placed over the window of the X-ray tube, filtering out the low energy end of the spectrum. This is called *hardening* the beam.

Both X-ray production processes are extremely inefficient (~1%) and thus to produce a usable flux of X-rays plenty of energy has to be wasted into heat, which has to be removed from the X-ray tube.

Radiographs obtained using X-rays can be used to identify a wide spectrum of pathologies. Due to their short wavelength, in medical applications, X-rays act more like a particle than a wave. This is in contrast to their application in crystallography, where their wave-like nature is most important.

To take an X-ray of the bones, short X-ray pulses are shot through a body with radiographic film behind. The bones absorb the most photons by the photoelectric process, because they are more electron-dense. The X-rays that do not get absorbed turn the photographic film from white to black, leaving a white shadow of bones on the film.

To generate an image of the cardiovascular system, including the arteries and veins (angiography) an initial image is taken of the anatomical region of interest.

A second image is then taken of the same region after iodinated contrast material has been injected into the blood vessels within this area. These two images are then digitally subtracted, leaving an image of only the iodinated contrast outlining the blood vessels.

The radiologist or surgeon then compares the image obtained to normal anatomical images to determine if there is any damage or blockage of the vessel.

A specialized source of X-rays which is becoming widely used in research is synchrotron radiation, which is generated by particle accelerators. Its unique features are brightness many orders of magnitude greater than X-ray tubes, wide spectrum, high collimation, and linear polarization.

DETECTORS

Photographic Plate

The detection of X-rays is based on various methods. The most commonly known methods are a photographic plate, X-ray film in a cassette, and rare earth screens. Regardless of what is "catching" the image, they are all categorized as "Image Receptors" (IR).

Before computers and before digital imaging, a photographic plate was used to produce radiographic images. The images were produced right on the glass plates. Film replaced these plates and was used in hospitals to produce images.

Now computed & digital radiography has started to replace film in medicine, though film technology remains in use in industrial radiography processes (e.g. to inspect welded seams).

Photographic plates are a thing of history, and their replacement (intensifying screens) is now becoming part of that same history. Silver (necessary to the radiographic & photographic industry) is a non-renewable resource, that has now been replaced by digital (DR) and computed (CR) technology. Where film required wet processing facilities, these new technologies do not. Archiving of these new technologies also saves space.

Since photographic plates are sensitive to X-rays, they provide a means of recording the image, but require a lot of exposure (to the patient), so intensifying screens were devised. They allow a lower dose to the patient, because the screens take the X-ray information and intensify it so that it can be recorded on film positioned next to the intensifying screen.

The part of the patient to be X-rayed is placed between the X-ray source and the image receptor to produce a shadow of the internal structure of that particular part of the body. X-rays are partially blocked ("attenuated") by dense tissues such as bone, and pass more easily through soft tissues. Areas where the X-rays strike darken when developed, causing bones to appear lighter than the surrounding soft tissue.

Contrast compounds containing barium or iodine, which are radiopaque, can be ingested in the gastrointestinal tract (barium) or injected in the artery or veins to highlight these vessels. The contrast compounds have high atomic numbered elements in them that (like bone) essentially block the X-rays and hence the once hollow organ or vessel can be more readily seen.

In the pursuit of a non-toxic contrast material, many types of high atomic number elements were evaluated. For example, the first time the forefathers used contrast it was chalk, and was used on a cadaver's vessels. Unfortunately, some elements chosen proved to be harmful – for example, thorium was once used as a contrast medium (Thorotrast) – which turned out to be toxic in some cases (causing injury and occasionally death from the effects of thorium poisoning).

Modern contrast material has improved, and while there is no way to determine who may have a sensitivity to the contrast, the incidence of "allergic-type reactions" are low. (The risk is comparable to that associated with penicillin.)

Photostimulable Phosphors (PSPs)

An increasingly common method of is the use of photostimulated luminescence (PSL), pioneered by Fuji in the 1980s. In modern hospitals a photostimulable phosphor plate (PSP plate) is used in place of the photographic plate.

After the plate is X-rayed, excited electrons in the phosphor material remain "trapped" in "colour centres" in the crystal lattice until stimulated by a laser beam passed over the plate surface.

The light given off during laser stimulation is collected by a photomultiplier tube and the resulting signal is converted into a digital image by computer technology, which gives this process its common name, computed radiography (also referred to as digital radiography). The PSP plate can be reused, and existing X-ray equipment requires no modification to use them.

Geiger Counter

Initially, most common detection methods were based on the ionization of gases, as in the Geiger-Müller counter: a sealed volume, usually a cylinder, with a mica, polymer or thin metal window contains a gas, and a wire, and a high voltage is applied between the cylinder (cathode) and the wire (anode).

When an X-ray photon enters the cylinder, it ionizes the gas and forms ions and electrons. Electrons accelerate toward the anode, in the process causing further ionization along their trajectory. This process, known as a Townsend avalanche, is detected as a sudden current, called a "count" or "event".

Ultimately, the electrons form a virtual cathode around the anode wire, drastically reducing the electric field in the outer portions of the tube. This halts the collisional ionizations and limits further growth of avalanches. As a result, all "counts" on a Geiger counter are the same size and it can give no indication as to the particle energy of the radiation, unlike the proportional counter. The intensity of the radiation is measurable by the Geiger counter as the counting-rate of the system.

In order to gain energy spectrum information, a diffracting crystal may be used to first separate the different photons. The method is called wavelength dispersive X-ray spectroscopy (WDX or WDS).

Position-sensitive detectors are often used in conjunction with dispersive elements. Other detection equipment that is inherently energy-resolving may be used, such as the aforementioned proportional counters. In either case, use of suitable pulse-processing (MCA) equipment allows digital spectra to be created for later analysis.

For many applications, counters are not sealed but are constantly fed with purified gas, thus reducing problems of contamination or gas aging. These are called "flow counters".

Scintillators

Some materials such as sodium iodide (NaI) can "convert" an X-ray photon to a visible photon; an electronic detector can be built by adding a photomultiplier.

These detectors are called "scintillators", filmscreens or "scintillation counters". The main advantage of using these is that an adequate image can be obtained while subjecting the patient to a much lower dose of X-rays.

Image Intensification

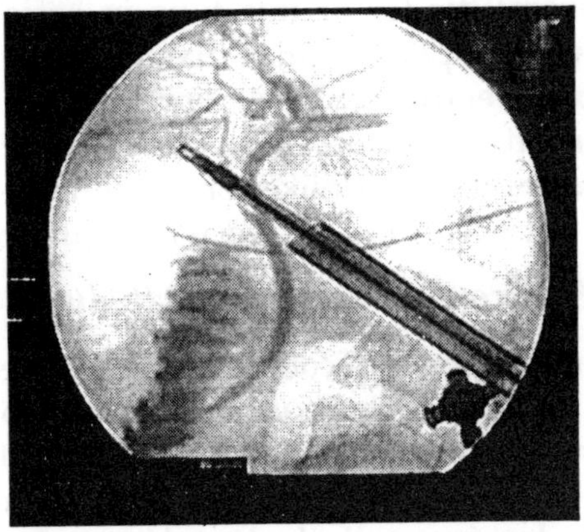

Fig. X-ray During Cholecystectomy.

X-rays are also used in "real-time" procedures such as angiography or contrast studies of the hollow organs (e.g. barium enema of the small or large intestine) using fluoroscopy acquired using **an X-ray image** intensifier. Angioplasty, medical interventions of the arterial system, rely heavily on X-ray-sensitive contrast to identify potentially treatable lesions.

Direct Semiconductor Detectors

Since the 1970s, new semiconductor detectors have been developed (silicon or germanium doped with lithium, Si(Li) or Ge(Li)). X-ray photons are converted to electron-hole pairs in the semiconductor and are collected to detect the X-rays.

When the temperature is low enough (the detector is cooled by Peltier effect or even cooler liquid nitrogen), it is possible to directly determine the X-ray energy spectrum; this method is called energy dispersive X-ray spectroscopy (EDX or EDS); it is often used in small X-ray fluorescence spectrometers. These detectors are sometimes called "solid state detectors". Cadmium telluride (CdTe) and its alloy with zinc, cadmium zinc telluride detectors have an increased sensitivity, which allows lower doses of X-rays to be used.

Practical application in medical imaging didn't start taking place until the 1990s. Currently amorphous selenium is used in commercial large area flat panel X-ray detectors for mammography and chest radiography. Current research and development is focused around pixel detectors, such as CERN's energy resolving Medipix detector.

Note: A standard semiconductor diode, such as a 1N4007, will produce a small amount of current when placed in an X-ray beam. A test device once used by Medical Imaging Service personnel was a small project box that contained several diodes of this type in series, which could be connected to an oscilloscope as a quick diagnostic.

Silicon drift detectors (SDDs), produced by conventional semiconductor fabrication, now provide a cost-effective and high resolving power radiation measurement. Unlike conventional X-ray detectors, such as Si(Li)s, they do not need to be cooled with liquid nitrogen.

Scintillator Plus Semiconductor Detectors (Indirect Detection)

With the advent of large semiconductor array detectors it has become possible to design detector systems using a scintillator screen to convert from X-rays to visible light which is then converted to electrical signals in an array detector. Indirect Flat Panel Detectors (FPDs) are in widespread use today in medical, dental, veterinary and industrial applications.

The array technology is a variant on the amorphous silicon TFT arrays used in many flat panel displays, like the ones in computer laptops. The array consists of a sheet of glass covered with a thin layer of silicon that is in an amorphous or disordered state. At a microscopic scale, the silicon has been imprinted with millions of transistors arranged in a highly ordered array, like the grid on a sheet of graph paper.

Each of these thin film transistors (TFTs) are attached to a light-absorbing photodiode making up an individual pixel (picture element). Photons striking the photodiode are converted into two carriers of electrical charge, called electron-hole pairs. Since the number of charge carriers produced will vary with the intensity of incoming light photons, an electrical pattern is created that can be swiftly converted to a voltage and then a digital signal, which is interpreted by a computer to produce a digital image.

Although silicon has outstanding electronic properties, it is not a particularly good absorber of X-ray photons. For this reason, X-rays first impinge upon scintillators made from e.g. gadolinium oxysulfide or caesium iodide. The scintillator absorbs the X-rays and converts them into visible light photons that then pass onto the photodiode array.

Visibility to the Human Eye

While generally considered invisible to the human eye, in special circumstances X-rays can be visible. Brandes, in an experiment a short time after Röntgen's landmark 1895 paper, reported after dark adaptation and placing his eye close to an X-ray tube, seeing a faint "blue-gray" glow which seemed to originate within the eye itself.

Upon hearing this, Röntgen reviewed his record books and found he too had seen the effect. When placing an X-ray tube on the opposite side of a wooden door Röntgen had noted the same blue glow, seeming to emanate from the eye itself, but thought his observations to be spurious because he only saw the effect when he used one type of tube.

Later he realized that the tube which had created the effect was the only one powerful enough to make the glow plainly visible and the experiment was thereafter readily repeatable. The knowledge that X-rays are actually faintly visible to the dark-adapted naked eye has largely been forgotten today; this is probably due to the desire not to repeat what would now be seen as a recklessly dangerous and potentially harmful experiment with ionizing radiation.

It is not known what exact mechanism in the eye produces the visibility: it could be due to conventional detection (excitation of rhodopsin molecules in the retina), direct excitation of retinal nerve cells, or secondary detection via, for instance, X-ray induction of phosphorescence in the eyeball with conventional retinal detection of the secondarily produced visible light.

Though X-rays are invisible it is possible to see the ionization of the air molecules if the intensity of the X-ray beam is high enough. The beamline from the wiggler at the ID11 at ESRF is one example of such high intensity.

Medical Uses

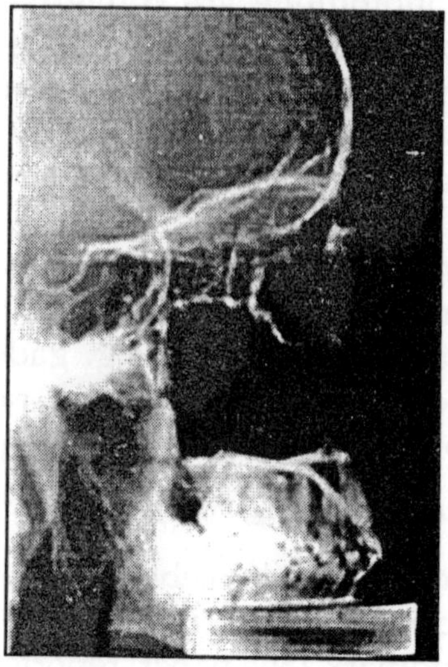

Fig. X-Ray Image of the Paranasal Sinuses, Lateral Projection.

Since Röntgen's discovery that X-rays can identify bony structures, X-rays have been developed for their use in medical imaging. Radiology is a specialized field of medicine. Radiologists employ radiography and other techniques for diagnostic imaging. This is probably the most common use of X-ray technology.

X-rays are especially useful in the detection of pathology of the skeletal system, but are also useful for detecting some disease processes in soft tissue. Some notable examples are the very common chest X-ray, which can be used to identify lung diseases such as pneumonia, lung cancer or pulmonary edema, and the abdominal X-ray, which can detect ileus (blockage of the intestine), free air (from visceral perforations) and free fluid (in ascites).

In some cases, the use of X-rays is debatable, such as gallstones (which are rarely radiopaque) or kidney stones (which are often visible, but not always).

Also, traditional plain X-rays pose very little use in the imaging of soft tissues such as the brain or muscle. Imaging alternatives for soft tissues are computed axial tomography (CAT or CT scanning), magnetic resonance imaging (MRI) or ultrasound. Since 2005, X-rays are listed as a carcinogen by the U.S. government.

Radiotherapy, a curative medical intervention, now used almost exclusively for cancer, employs higher energies of radiation.

Shielding Against X-Rays

Lead is the most common shield agains X-Rays because of its high density (11340 kg/m^3), its easiness of installation and low price.

The following table shows the recommended thickness of lead shielding in function of X-Ray energy, from the Recommendations by the Second International Congress of Radiology.

X-Rays generated by peak voltagesnot exceeding	Minimum thicknessof Lead
75 kV	1.0 mm
100 kV	1.5 mm
125 kV	2.0 mm
150 kV	2.5 mm
175 kV	3.0 mm
200 kV	4.0 mm
225 kV	5.0 mm
300 kV	9.0 mm
400 kV	15.0 mm
500 kV	22.0 mm
600 kV	34.0 mm
900 kV	51.0 mm

Other Uses

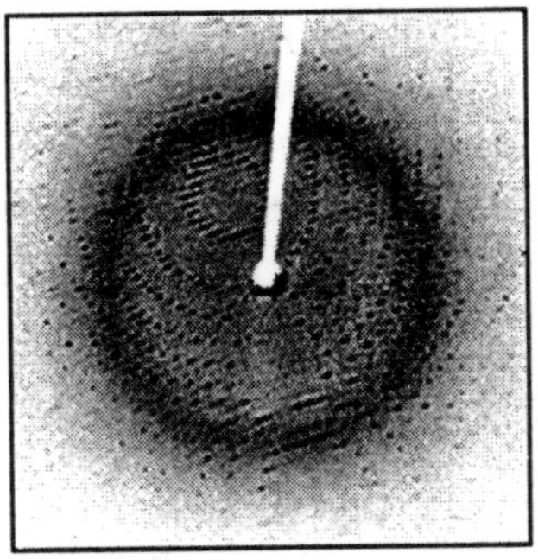

Fig. Diffraction Pattern

Each dot, called a reflection, in this diffraction pattern forms from the constructive interference of scattered X-rays passing through a crystal. The data can be used to determine the crystalline structure.

Other notable uses of X-rays include:

- X-ray crystallography in which the pattern produced by the diffraction of X-rays through the closely spaced lattice of atoms in a crystal is recorded and then analysed to reveal the nature of that lattice. A related technique, fibre diffraction, was used by Rosalind Franklin to discover the double helical structure of DNA.
- X-ray astronomy, which is an observational branch of astronomy, which deals with the study of X-ray emission from celestial objects.
- X-ray microscopic analysis, which uses electromagnetic radiation in the soft X-ray band to produce images of very small objects.
- X-ray fluorescence, a technique in which X-rays are generated within a specimen and detected. The outgoing energy of the X-ray can be used to identify the composition of the sample.
- Industrial radiography uses X-rays for inspection of industrial parts, particularly welds.
- Paintings are often X-rayed to reveal the underdrawing and pentimenti or alterations in the course of painting, or by later restorers. Many pigments such as lead white show well in X-ray photographs.
- Airport security luggage scanners use X-rays for inspecting the interior of luggage for security threats before loading on aircraft.

X-rays were discovered emanating from Crookes tubes, experimental discharge tubes invented around 1875, by scientists investigating the cathode rays, that is energetic electron beams, that were first created in the tubes.

Crookes tubes created electrons by ionization of the residual air in the tube by a high DC voltage of anywhere between a few kilovolts and 100 kV. This voltage accelerated the electrons coming from the cathode to a high enough velocity that they created X-rays when they struck the anode or the glass wall of the tube.

Many of the early Crookes tubes undoubtedly radiated X-rays, because early researchers noticed effects that were attributable to them, as detailed below, before Wilhelm Röntgen first systematically studied them in 1895. Among the important early researchers in X-rays were Ivan Pulyui, William Crookes, Johann Wilhelm Hittorf, Eugen Goldstein, Heinrich Hertz, Philipp Lenard, Hermann von Helmholtz, Nikola Tesla, Thomas Edison, Charles Glover Barkla, Max von Laue, and Wilhelm Conrad Röntgen.

Wilhelm Röntgen

On November 8, 1895, German physics professor Wilhelm Conrad Röntgen, stumbled on X-rays while experimenting with Lenard and

Crookes tubes and began studying them. He wrote an initial report "*On a new kind of ray: A preliminary communication*" and on December 28, 1895 submitted it to the Würzburg's Physical-Medical Society journal.

This was the first paper written on X-rays. Röntgen referred to the radiation as "X", to indicate that it was an unknown type of radiation. The name stuck, although (over Röntgen's great objections), many of his colleagues suggested calling them Röntgen rays. They are still referred to as such in many languages, including German. Röntgen received the first Nobel Prize in Physics for his discovery.

There are conflicting accounts of his discovery because Röntgen had his lab notes burned after his death, but this is a likely reconstruction by his biographers: Röntgen was investigating cathode rays with a fluorescent screen painted with barium platinocyanide and a Crookes tube which he had wrapped in black cardboard so the visible light from the tube wouldn't interfere.

He noticed a faint green glow from the screen, about 1 meter away. The invisible rays coming from the tube to make the screen glow were passing through the cardboard. He found they could also pass through books and papers on his desk. Röntgen threw himself into investigating these unknown rays systematically. Two months after his initial discovery, he published his paper.

Röntgen discovered its medical use when he saw a picture of his wife's hand on a photographic plate formed due to X-rays. His wife's hand's photograph was the first ever photograph of a human body part using X-rays.

Johann Hittorf

German physicist Johann Hittorf, a coinventor and early researcher of the Crookes tube, found when he placed unexposed photographic plates near the tube, that some of them were flawed by shadows, though he did not investigate this effect.

Ivan Pulyui

In 1877 Ukranian-born Pulyui, a lecturer in experimental physics at the University of Vienna, constructed various designs of vacuum discharge tube to investigate their properties. He continued his investigations when appointed professor at the Prague Polytechnic and in 1886 he found that that sealed photographic plates became dark when exposed to the emanations from the tubes.

Early in 1896, just a few weeks after Röntgen published his first X-ray photograph, Pulyui published high-quality x-ray images in journals in Paris and London. Although Pulyui had studied with Röntgen at the

University of Strasbourg in the years 1873-75, his biographer Gaida asserts that his subsequent research was conducted independently.

The first medical X-ray made in the United States was obtained using a discharge tube of Pulyui's design. In January 1896, on reading of Röntgen's discovery, Frank Austin of Dartmouth College tested all of the discharge tubes in the physics laboratory and found that only the Pulyui tube produced X-rays. This was a result of Pulyui's inclusion of an oblique "target" of mica, used for holding samples of fluorescent material, within the tube.

On 3 February 1896 Gilman Frost, professor of medicine at the college, and his brother Edwin Frost, professor of physics, exposed the wrist of Eddie McCarthy, whom Edwin had treated some weeks earlier for a fracture, to the X-rays and collected the resulting image of the broken bone on gelatin photographic plates obtained from Howard Langill, a local photographer also interested in Röntgen's work.

Nikola Tesla

In April 1887, Nikola Tesla began to investigate X-rays using high voltages and tubes of his own design, as well as Crookes tubes. From his technical publications, it is indicated that he invented and developed a special single-electrode X-ray tube, which differed from other X-ray tubes in having no target electrode.

The principle behind Tesla's device is called the Bremsstrahlung process, in which a high-energy secondary X-ray emission is produced when charged particles (such as electrons) pass through matter.

By 1892, Tesla performed several such experiments, but he did not categorize the emissions as what were later called X-rays. Tesla generalized the phenomenon as radiant energy of "invisible" kinds.

Tesla stated the facts of his methods concerning various experiments in his 1897 X-ray lecture before the New York Academy of Sciences. Also in this lecture, Tesla stated the method of construction and safe operation of X-ray equipment. His X-ray experimentation by vacuum high field emissions also led him to alert the scientific community to the biological hazards associated with X-ray exposure.

Fernando Sanford

X-rays were generated and detected by Fernando Sanford (1854-1948), the foundation Professor of Physics at Stanford University, in 1891.

From 1886 to 1888 he had studied in the Hermann Helmholtz laboratory in Berlin, where he became familiar with the cathode rays generated in vacuum tubes when a voltage was applied across separate electrodes, as previously studied by Heinrich Hertz and Philipp Lenard.

His letter of January 6, 1893 (describing his discovery as "electric photography") to The Physical Review was duly published and an article entitled *Without Lens or Light, Photographs Taken With Plate and Object in Darkness* appeared in the San Francisco Examiner.

Philipp Lenard

Philipp Lenard, a student of Heinrich Hertz, wanted to see whether cathode rays could pass out of the Crookes tube into the air. He built a Crookes tube (later called a "Lenard tube") with a "window" in the end made of thin aluminum, facing the cathode so the cathode rays would strike it. He found that something came through, that would expose photographic plates and cause fluorescence.

He measured the penetrating power of these rays through various materials. It has been suggested that at least some of these "Lenard rays" were actually X-rays. Hermann von Helmholtz formulated mathematical equations for X-rays. He postulated a dispersion theory before Röntgen made his discovery and announcement. It was formed on the basis of the electromagnetic theory of light. However, he did not work with actual X-rays.

Thomas Edison

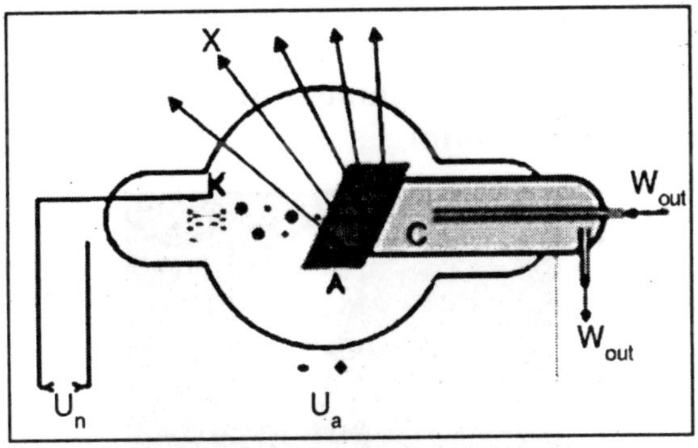

Fig. Diagram of a Water Cooled X-ray Tube.

In 1895, Thomas Edison investigated materials' ability to fluoresce when exposed to X-rays, and found that calcium tungstate was the most effective substance. Around March 1896, the fluoroscope he developed became the standard for medical X-ray examinations.

Nevertheless, Edison dropped X-ray research around 1903 after the death of Clarence Madison Dally, one of his glassblowers. Dally had a habit of testing X-ray tubes on his hands, and acquired a cancer in them so tenacious that both arms were amputated in a futile attempt to save his life.

"At the 1901 Pan-American Exposition in Buffalo, New York, an assassin shot President William McKinley twice at close range with a .32 caliber revolver." The first bullet was removed but the second remained lodged somewhere in his stomach. McKinley survived for some time and requested that Thomas Edison "rush an X-ray machine to Buffalo to find the stray bullet. It arrived *but wasn't used*. McKinley died of septic shock due to bacterial infection."

The 20th Century and Beyond

The many applications of X-rays immediately generated enormous interest. Workshops began making specialized versions of Crookes tubes for generating X-rays, and these first generation cold cathode or Crookes X-ray tubes were used until about 1920.

Crookes tubes were unreliable. They had to contain a small quantity of gas (invariably air) as a current will not flow in such a tube if they are fully evacuated. However as time passed the X-rays caused the glass to absorb the gas, causing the tube to generate "harder" X-rays until it soon stopped operating. Larger and more frequently used tubes were provided with devices for restoring the air, known as "softeners".

This often took the form of a small side tube which contained a small piece of mica – a substance that traps comparatively large quantities of air within its structure. A small electrical heater heated the mica and caused it to release a small amount of air restoring the tube's efficiency. However the mica itself had a limited life and the restore process was consequently difficult to control.

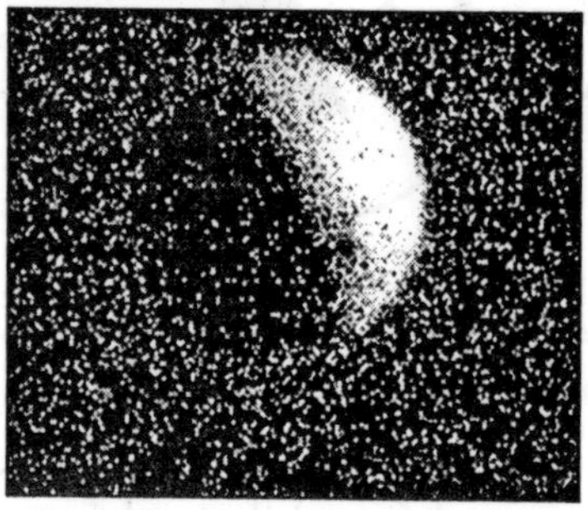

Fig. ROSAT Image of X-ray Fluorescence of, and Occultation of the X-ray Background by, the Moon.

In 1904, John Ambrose Fleming invented the thermionic diode valve (vacuum tube). This used a hot cathode which permitted current to flow in a vacuum. This idea was quickly applied X-ray tubes, and heated cathode X-ray tubes, called Coolidge tubes, replaced the troublesome cold cathode tubes by about 1920.

Two years later, physicist Charles Barkla discovered that X-rays could be scattered by gases, and that each element had a characteristic X-ray. He won the 1917 Nobel Prize in Physics for this discovery. Max von Laue, Paul Knipping and Walter Friedrich observed for the first time the diffraction of X-rays by crystals in 1912.

This discovery, along with the early works of Paul Peter Ewald, William Henry Bragg and William Lawrence Bragg gave birth to the field of X-ray crystallography. The Coolidge tube was invented the following year by William D. Coolidge which permitted continuous production of X-rays; this type of tube is still in use today.

The use of X-rays for medical purposes (to develop into the field of radiation therapy) was pioneered by Major John Hall-Edwards in Birmingham, England. In 1908, he had to have his left arm amputated owing to the spread of X-ray dermatitis.

The X-ray microscope was invented in the 1950s.

The Chandra X-ray Observatory, launched on July 23, 1999, has been allowing the exploration of the very violent processes in the universe which produce X-rays. Unlike visible light, which is a relatively stable view of the universe, the X-ray universe is unstable, it features stars being torn apart by black holes, galactic collisions, and novas, neutron stars that build up layers of plasma that then explode into space.

An X-ray laser device was proposed as part of the Reagan Administration's Strategic Defence Initiative in the 1980s, but the first and only test of the device (a sort of laser "blaster", or death ray, powered by a thermonuclear explosion) gave inconclusive results.

For technical and political reasons, the overall project (including the X-ray laser) was de-funded (though was later revived by the second Bush Administration as National Missile Defence using different technologies).

Bibliography

Asomoza, M: *Hydrolysis catalyst effect on sol-gel silica structure*, 1999.

Bates, S: "*Assessment of Defects and Amorphous Structure produced in Raffinose Pentahydrate upon dehydration*", 1995.

Bates, S: "*Effect of Water-Vapor Sorption on Local Structure of Polyvinylpyrrolidone*", 1999.

Bethune, S: "*Nutraceutical cocrystals: utilizing pterostilbene as a cocrystal former*", 1996.

Catlow, R: *Solid state chemistry: New opportunities from computer simulations*, 2000.

Chen, P: "*A novel method for the assessment of miscibility in amorphous dispersions*", 1996.

Coltrain, B. K: *The Chemistry of Hydrolysis and Condensation of Silica Sol-Gel Precursors*, 1996.

Damrau, U: *The hydrolysis of oligomer intermediates in the sol-gel process*, 1998.

Dunlap, J: "*New Solid-State Chemistry Technologies to Bring Better Drugs to Market: Knowledge Based Decision Making*", 1995.

Dunuwila, D: *Application of Controlled Hydrolysis* of Titanium, 2002.

Hanko, J: "*Three isostructural solvates of finasteride and their solid-state characterization*", 1995.

Hardcastle, K. I: "*Cocrystals of Piroxicam with Carboxylic Acids*", 1999.

Hilfiker, R: *Polymorphism in the pharmaceutical industry*, 1998.

Ivanisevic, I: "*Physical Stability Studies of Miscible Amorphous Solid Dispersions*", 1998.

Jones, W: "*The role of solvent in mechanochemical and sonochemical cocrystal formation: a solubility-based approach for predicting cocrystallisation outcome*", 1999.

Joshi, V: *The influence of water of hydrolysis on microstructural development in sol-gel derived LiNbO3 thin films*, 1999.

Ladik, J: *Quantum chemistry of polymers—solid state aspects*, 1995.

Maheshwari, C: "*Screening strategies based on solubility and solution composition generate pharmaceutically acceptable cocrystals of carbamazepine*", 1995.

Newman, A: "*Characterization of amorphous API:Polymer mixtures using X-ray powder diffraction*", 2000.

Pikal, M.J: "*The Study of Phase Separation in Amorphous Freeze-Dried Systems, Part I: Raman Mapping and Computational Analysis of XRPD Data in Model Polymer Systems*", 1999.

Roe, M: "*Cocrystals of nutraceutical p-coumaric acid with caffeine and theophylline: polymorphism and solid-state stability explored in detail using their crystal graphs*", 1995.

Roger, C: *Formation of porous metal oxides via sol-gel type hydrolysis of metal alkoxide complexes modified with organic templates*, 1998.

Rumondor, A.C.F: "*Evaluation of Drug-Polymer Miscibility in Amorphous Solid Dispersion Systems*", 2000.

Stahly, G. P: "*Discovery of a New Polymorph of Dehydroepiandrosterone (Prasterone) and Solution of Its Crystal Structure for X-ray Powder Diffraction Data*", 1999.

Stanitski, L: *Chemistry in Context: Applying Chemistry to Society*, 1999.

Stout, H: *X-Ray Structure Determination*, 1996.

Teng, J: "*A Solid-State Approach to Enable Early Development Compounds: Selection and Animal Bioavailability Studies of an Itraconazole Amorphous Solid Dispersion*", 1998.

Wade, R: *Surface Raman Scattering of Self-Assembled Monolayers*, 1999.

Yeager, E: *Magnesium Methoxide Derived Catalysis of Transesterification, Hydrolysis, and Condensation of Si(OMe)4 in the Sol-Gel Syntheses of Magnesium Silicates*, 1999.

Index

F

G

H

I

L

M

N

O

P

Q

R

S

T

W

X